高等学校“十三五”规划教材

Modern Control Theory

现代控制理论

（中英文对照）

李道根　主编

哈爾濱工業大學出版社

Brief of contents

This book states the elementary principle of modern control theory and discusses its analysis and synthesis techniques based on "state space method". The book is divided in five chapters, including state space description of linear systems, dynamic analysis of linear systems, controllability and observability of linear systems, state-space synthesis of linear systems, and Lyapunov stability analysis.

This book is a product of the bilingual teaching practice and experience of the Editors in Jiangsu University of Science and Technology. It can be used as a textbook of automation, electrician engineering and automation specialties, and so on, as well as a reference book for relevant engineers.

内容简介

全书以状态空间法为基础阐述了现代控制理论的基本原理及其分析和综合方法。全书分五章,内容包括线性系统的状态空间描述、线性系统的运动分析、线性系统的能控性和能观测性、线性系统的状态综合以及李雅普诺夫稳定性分析。

本书是在江苏科技大学近年来现代控制理论双语教学的基础上编写的。本书可作为高等院校自动化、电气工程及其自动化等控制类专业的双语教学教材,也可作为相关工程技术人员的参考书。

图书在版编目(CIP)数据

现代控制理论(中英文对照)/李道根主编.—哈尔滨:哈尔滨工业大学出版社,2009.3(2018.7 重印)
ISBN 978-7-5603-2811-9

Ⅰ.现… Ⅱ.李… Ⅲ.现代控制理论-汉、英
Ⅳ.O231

中国版本图书馆 CIP 数据核字(2009)第 023247 号

策划编辑 张秀华
责任编辑 范业婷
封面设计 卞秉利
出版发行 哈尔滨工业大学出版社
社　　址 哈尔滨市南岗区复华四道街 10 号　邮编 150006
传　　真 0451-86414749
网　　址 http://hitpress.hit.edu.cn
印　　刷 哈尔滨市工大节能印刷厂
开　　本 787mm×1092mm　1/16　印张 11.75　字数 267 千字
版　　次 2009 年 3 月第 1 版　2018 年 7 月第 3 次印刷
书　　号 ISBN 978-7-5603-2811-9
定　　价 24.00 元

Preface

Bilingual teaching is the inevitable trend of education reform, also the important action to improve undergraduate education quality, and the requirement of cultivate compound-type talent with high quality who can adapt social development in the 21-the century. Finding one adapted way to ordinary colleges will be helpful to enhance the student's ability of English application and understanding of the latest scientific achievements.

National Ministry of Education calls for actively promotion to develop bilingual teaching in colleges and universities, early of 2007, National Ministry of Education and Ministry of Finance jointly introduced the "Opinion on the implementation of quality of undergraduate education and teaching reform", encouraging colleges and universities to build 500 bilingual teaching model curriculums within five years.

So it is significant to research the special feature, method and mode of bilingual teaching and improve level of special English of students as soon as possible. The compilers of this book conforms to the trend of the development of higher education, carried through bilingual teaching of "Automatic control principle" and "Modern control theory" in automation and electrical engineering and automation specialties in Jiangsu university of science and technology. This book is one of the achievements of their bilingual teaching practice.

In despite of these or those difficulties, bilingual teaching will be developed rapidly with the update of education idea and make contribution to enhance the bilingual ability and compositive human culture quality of students.

Wang Jianhua
February, 2009

序

双语教学是我国高等教育改革发展的必然趋势，是提高我国本科教学质量的重要举措，也是培养适应21世纪社会发展高素质复合型人才的需要。寻找到一套适合普通高校双语教学的方法，对于提高学生的英语实际应用能力和了解世界科技最新成果的能力，无疑会具有很大的帮助。

为此，教育部要求各高校积极推动使用英语等外语进行公共课和专业课的教学。2007年初，教育部和财政部又联合出台了《关于实施高等学校本科教学质量与教学改革工程的意见》，提出在五年内要建设500门双语教学示范课程。

因此，研究“双语”教学的特点，积极探索“双语”教学的教学方法、教学模式，以尽快提高学生的专业英语水平，对教学研究具有十分重要的意义。编著者顺应高等教育的发展潮流，对江苏科技大学的自动化和电气工程及其自动化专业的“自动控制原理”和“现代控制理论”课程进行了双语教学，本书正是编著者开展双语教学研究的成果之一。

尽管双语教学在我国还存在着这样或那样的问题，但随着我国教育理念的更新，双语教学必定会取得长足的进展，为提高学生的“双语”能力和综合人文素质作出贡献。

王建华

2009年2月

Foreword

As one of the basic curricula of control specialties, "Modern control theory" is significant for studying other subjects of control theory, such as Optimal control, System identification, Robust control and Intelligent control, and it is also widely used in so many fields, such as industry, agriculture, shipbuilding and aviation.

In recently years the Compilers carried through bilingual teaching of "Modern control theory" in automation and electrical engineering and automation specialties, this book is a product of our bilingual teaching practice and experience. The book takes bilingual form to provide convenience for readers.

Based on "state space method", this book systematically expatiates the elementary theory and methods of linear systems. We try our best to explain the deep things in a simple way and apply rigorous theory. We expect this book would play an active role in improving English application ability of the readers, cultivating correction ideation and the ability to connect theory with practice of students.

This book can be used as a professional teaching material of automation, electrician engineering and automation specialties, and so on, as well as a reference book for graduate students and teachers.

This book is edited by Li Daogen. Dr. Zhu Zhiyu and Liu Weiting make contribution to this book. We would especially like to thank Hao Peng, Wang Fang and Zong Yang for drawing figures and tables in this book. This book is reviewed by Professor Jiang Changsheng. And we would like to thank all the teachers and students for their help in the production and editing of this book.

Some mistakes are inevitable in this book because of our limited level, any criticizing correction will be appreciated.

Compiler
February, 2009

前　言

“现代控制理论”是控制类专业的一门专业基础课程，本门课程对于学习控制理论的许多学科分支，如最优控制、系统辨识、鲁棒控制、智能控制等具有重要的作用，同时该理论也广泛应用于工业、农业、船舶、航空航天等诸多领域。

近几年来，编者对自动化和电气工程及其自动化等专业的“现代控制理论”课程进行了双语教学尝试，本书是编者在开展“双语”教学实践的基础上，结合多年讲授“现代控制理论”课程的心得和体会，并在参考了大量原版教材的基础上编写而成的。全书采用中英文对照的形式出版，以方便读者阅读和学习。

全书以状态空间描述法为核心，系统地阐述了线性系统的基本控制理论和方法。全书力求做到深入浅出，理论严谨。本书对提高读者的英语应用能力，培养学生的辩证思维能力和理论联系实际的能力，都具有一定的作用。

本书可用于工科自动化、电气工程及其自动化专业的教材，也可作为非本专业研究生和教师的参考书。

本书由李道根主编，朱志宇、刘维亭等参与了部分编写工作，郝鹏、王芳、宗阳等同学绘制了书中的图表。全书由南京航空航天大学姜长生教授审定。在此，向曾参与和关心本书编写工作的各位教师和同学表示感谢。

作者水平有限，对于书中存在的错误和不妥之处，敬请读者批评指正。

编　者

2009 年 2 月

Contents

目 录

Chapter 1 State-Space Description of Linear Systems

1.1 Introduction

Nowadays, one of the challenges to control engineers is the modeling and control of modern, complex, large-scale and interrelated systems, such as chemical processes, robotic systems, and traffic control systems. Although, in classical control theory, the transfer function description of physical systems provides a practical approach to analysis and design and allows us to use block diagrams to interconnect subsystems, it has certain basic limitations. This is due to the fact that it is an external description of control systems based on the input-output relation, and is only applicable to the linear, time-invariant single-variable systems, i.e. systems with one input and one output only.

On the other hand, the time-domain method based on the state-space description is more powerful. Its principle advantages are that it is easily implemented on a digital computer, it is applicable to nonlinear systems and time-varying systems, it is readily extended to multivariable systems, i.e. systems with several inputs and outputs, and it provides additional insight into system behavior that transfer function analysis does not.

1.1.1 State Variable

The basis of modern control theory is the concepts of state and state variable. As the word implies, the state of a system refers to the past, present, and future conditions of the system. In general, the state can be described by a set of numbers, a curve, an equation, or something that is more abstract in nature. For dynamic system, however, the state of a system is described in terms of a set of state variables.

The state variables of a dynamic system is defined as the minimum set of variables such that the initial values of these variables at time t_0, together with any input to the system for $t \geqslant t_0$, will allow the behavior of the system to be determined uniquely for $t \geqslant t_0$. It should be emphasized, however, that the selection of state variables is not unique; there are many different sets of state variables that can be available for a given system.

A simple example of the state variable description of a dynamic system is the RLC circuit shown in Fig. 1.1. For a passive RLC network, the number of state variables required is equal to the number of independent energy-storage elements. The state of this system can be described in terms of a set of state variables: x_1, x_2, where x_1 is the inductor current $i_L(t)$ and x_2 is the capacitor voltage $v_C(t)$. This choice of state variables is intuitively satisfactory because the stored energy of the network can be described in terms of these variables as

$$E = \frac{1}{2}Li_L^2 + \frac{1}{2}Cv_C^2$$

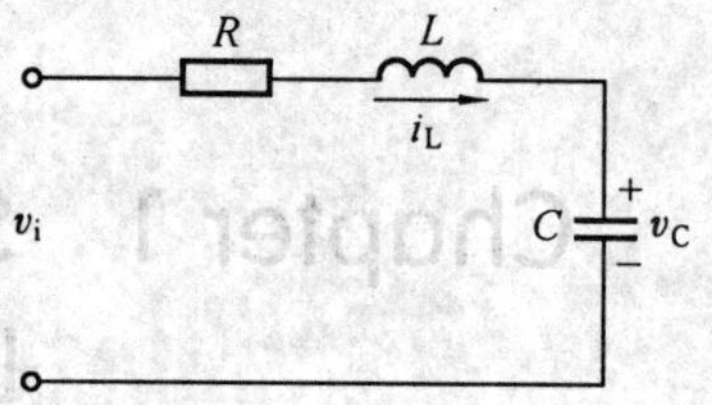

Figure 1.1 State variable model of RLC network

Using Kirchhoff's circuit laws, we obtain following equations describing the change rate of capacitor voltage and that of inductor current, respectively

$$L\frac{di_L}{dt} = -Ri_L - v_C + v_i$$

$$C\frac{dv_C}{dt} = i_L$$

We can rewrite these equations as a set of first-order differential equations in terms of the state variables x_1 and x_2 as follows

$$\frac{dx_1}{dt} = -\frac{R}{L}x_1 - \frac{1}{L}x_2 + \frac{1}{L}v_i$$

$$\frac{dx_2}{dt} = \frac{1}{C}x_1$$

Using above equations and the initial conditions of the network represented by $x_1(t_0)$ and $x_2(t_0)$, we can determine the system's future behavior.

For a particular system, the number of state variables is fixed and is equal to the system order. Or, we can say that the number of state variables may be determined as the number of initial conditions needed to solve the differential equation or the number of first-order differential equations needed to define the system.

1.1.2 State-space Representation

The general form of a dynamic system is shown in Fig. 1.2.

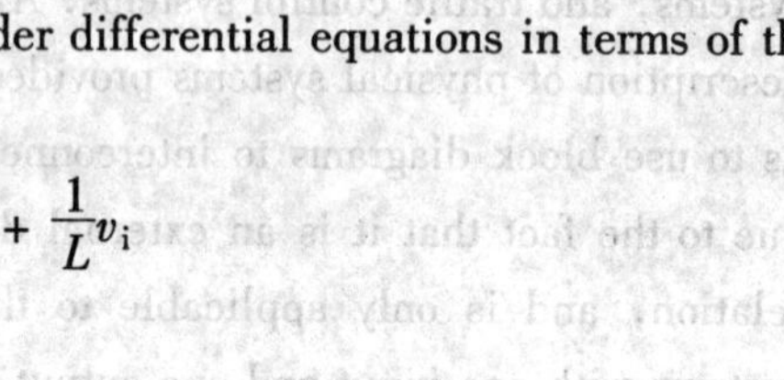

Figure 1.2 Dynamic system described with state variables

Since the system behavior can be completely described with state variables, the state of a nth-order linear time-invariant system can be described by a set of n first-order differential equations, written in terms of the state variables $x_1, x_2, \cdots, x_n$, in general form, as

$$\left.\begin{aligned} \dot{x}_1(t) &= a_{11}x_1(t) + a_{12}x_2(t) + \cdots + a_{1n}x_n(t) + b_{11}u_1(t) + \cdots + b_{1p}u_p(t) \\ \dot{x}_2(t) &= a_{21}x_1(t) + a_{22}x_2(t) + \cdots + a_{2n}x_n(t) + b_{21}u_1(t) + \cdots + b_{2p}u_p(t) \\ &\vdots \\ \dot{x}_n(t) &= a_{n1}x_1(t) + a_{n2}x_2(t) + \cdots + a_{nn}x_n(t) + b_{n1}u_1(t) + \cdots + b_{np}u_p(t) \end{aligned}\right\} \quad (1.1)$$

where $\dot{x} = \frac{dx}{dt}$. This set of differential equations can be written in the matrix form as follows

$$\begin{bmatrix} \dot{x}_1 \\ \dot{x}_2 \\ \vdots \\ \dot{x}_n \end{bmatrix} = \begin{bmatrix} a_{11} & a_{12} & \cdots & a_{1n} \\ a_{21} & a_{22} & \cdots & a_{2n} \\ \vdots & \vdots & & \vdots \\ a_{n1} & a_{n2} & \cdots & a_{nn} \end{bmatrix} \begin{bmatrix} x_1 \\ x_2 \\ \vdots \\ x_n \end{bmatrix} + \begin{bmatrix} b_{11} & b_{12} & \cdots & b_{1p} \\ b_{21} & b_{22} & \cdots & b_{2p} \\ \vdots & \vdots & & \vdots \\ b_{n1} & b_{n2} & \cdots & b_{np} \end{bmatrix} \begin{bmatrix} u_1 \\ u_2 \\ \vdots \\ u_p \end{bmatrix} \tag{1.2}$$

The column vector consisting the state variables is called a state vector and is written as

$$\boldsymbol{x} = \begin{bmatrix} x_1 \\ x_2 \\ \vdots \\ x_n \end{bmatrix}$$

The vector of input signals is denoted as $\boldsymbol{u}$. Then the system can be represented with the vector-matrix form of the state differential equation as

$$\dot{\boldsymbol{x}}(t) = \boldsymbol{A}\boldsymbol{x}(t) + \boldsymbol{B}\boldsymbol{u}(t) \tag{1.3}$$

Eq. (1.3) is also commonly called the state equation. The $n \times n$ square matrix $\boldsymbol{A}$ is called the system matrix; the $n \times p$ matrix $\boldsymbol{B}$ is called the input matrix. In this text, bold uppercase characters refer to matrices, while bold lowercase characters refer to vectors.

In general, the outputs of a linear system can be related to the state variables and the input signals by the output equation

$$\begin{bmatrix} y_1 \\ y_2 \\ \vdots \\ y_q \end{bmatrix} = \begin{bmatrix} c_{11} & c_{12} & \cdots & c_{1n} \\ c_{21} & c_{22} & \cdots & c_{2n} \\ \vdots & \vdots & & \vdots \\ c_{q1} & c_{q2} & \cdots & c_{qn} \end{bmatrix} \begin{bmatrix} x_1 \\ x_2 \\ \vdots \\ x_n \end{bmatrix} + \begin{bmatrix} d_{11} & d_{12} & \cdots & d_{1p} \\ d_{21} & d_{22} & \cdots & d_{2p} \\ \vdots & \vdots & & \vdots \\ d_{q1} & d_{q2} & \cdots & d_{qp} \end{bmatrix} \begin{bmatrix} u_1 \\ u_2 \\ \vdots \\ u_p \end{bmatrix}$$

i.e.

$$\boldsymbol{y}(t) = \boldsymbol{C}\boldsymbol{x}(t) + \boldsymbol{D}\boldsymbol{u}(t) \tag{1.4}$$

where $\boldsymbol{y}$ is the set of output signals expressed in column vector form, the $q \times n$ matrix $\boldsymbol{C}$ is called the output matrix, and the $q \times p$ matrix $\boldsymbol{D}$ is called the feedforward matrix or direct transfer matrix. In many practical examples we have $\boldsymbol{D} = \boldsymbol{0}$.

The combination of state-equation and output equation is called the state-space representation or state-space description. The block diagram shown in Fig. 1.3 is helpful for visualizing the relationship between the various elements and signals. The double lines represent vectors, rather than single variable, and the integral block symbolizes n integrators, one for each state variable.

As for the linear time-varying systems, the standard state-space representation is of the form

$$\dot{\boldsymbol{x}}(t) = \boldsymbol{A}(t)\boldsymbol{x}(t) + \boldsymbol{B}(t)\boldsymbol{u}(t)$$
$$\boldsymbol{y}(t) = \boldsymbol{C}(t)\boldsymbol{x}(t) + \boldsymbol{D}(t)\boldsymbol{u}(t)$$

where the matrixes $\boldsymbol{A}(t)$, $\boldsymbol{B}(t)$, $\boldsymbol{C}(t)$, and $\boldsymbol{D}(t)$ are continuous functions of time.

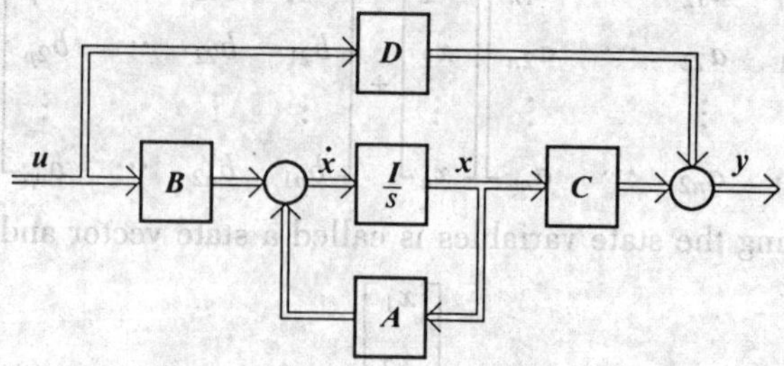

Figure 1.3 Block diagram of general state-space description

1.2 State-Space Representation from Input-Output Representation

In practice, it is often required to establish the space-state representation from the system input-output representation, i.e. differential equation describing system's dynamics. For linear time-invariant systems it is possible to develop the same result from the corresponding transfer function. In this section, we will consider only linear time-invariant single-variable systems, so the inputs and outputs will be scalars. Since the state variables of a given system are not unique, in general, we prefer to establish some "special" or canonical forms.

1.2.1 Input without Differential Items

Generally, a nth-order linear time-invariant system, for input without differential items, can be described by a differential equation

$$y^{(n)}(t) + a_1 y^{(n-1)}(t) + \cdots + a_{n-1}\dot{y}(t) + a_n y(t) = u(t) \tag{1.5}$$

where $y(t)$ is the output variable and $u(t)$ is the input. The corresponding transfer function is given by

$$\frac{Y(s)}{U(s)} = \frac{1}{s^n + a_1 s^{n-1} + \cdots + a_{n-1}s + a_n} \tag{1.6}$$

1. Controllable Canonical Form

For the present case, we define the state variables as

$$\left.\begin{aligned} x_1(t) &= y(t) \\ x_2(t) &= \dot{y}(t) \\ &\vdots \\ x_n(t) &= y^{(n-1)}(t) \end{aligned}\right\} \tag{1.7}$$

Then, we can obtain a set of first-order differential equations

$$\left.\begin{aligned} \dot{x}_1(t) &= x_2(t) \\ \dot{x}_2(t) &= x_3(t) \\ &\vdots \\ \dot{x}_n(t) &= -a_n x_1(t) - a_{n-1}x_2(t) - \cdots - a_1 x_n(t) + u(t) \end{aligned}\right\} \tag{1.8}$$

In Eq. (1.8) the last equation is obtained by equating the highest-ordered derivative term to the rest of Eq. (1.5). The output equation is simply

$$y(t) = x_1(t) \tag{1.9}$$

In the vector-matrix form, the state-space representation is written as

$$\left.\begin{aligned} \dot{\boldsymbol{x}}(t) &= \left[\begin{array}{c:ccc} 0 & & & \\ \vdots & & \boldsymbol{I}_{n-1} & \\ 0 & & & \\ \hdashline -a_n & -a_{n-1} & \cdots & -a_1 \end{array}\right]\boldsymbol{x}(t) + \begin{bmatrix} 0 \\ \vdots \\ 0 \\ 1 \end{bmatrix} u(t) \\ y(t) &= [1 \quad 0 \quad \cdots \quad 0]\boldsymbol{x}(t) \end{aligned}\right\} \tag{1.10}$$

For reasons to be discussed later, a state-space representation is said to be in the controllable canonical form if it has such a pair of system matrix and input matrix as shown in Eq. (1.10). Note that the system matrix is a companion matrix, and the coefficients that appear in the differential equation or transfer function also appear directly in the system matrix. In fact, it is evident that this canonical form can be written directly from the given differential equation or transfer function.

The block diagram shown in Fig. 1.4 is drawn according to the state-space representation given by Eq. (1.10), and is called the state variable diagram. This diagram is consists of integrators, multiplicators, summators, and signal lines, and the output of each integrator is a state variable.

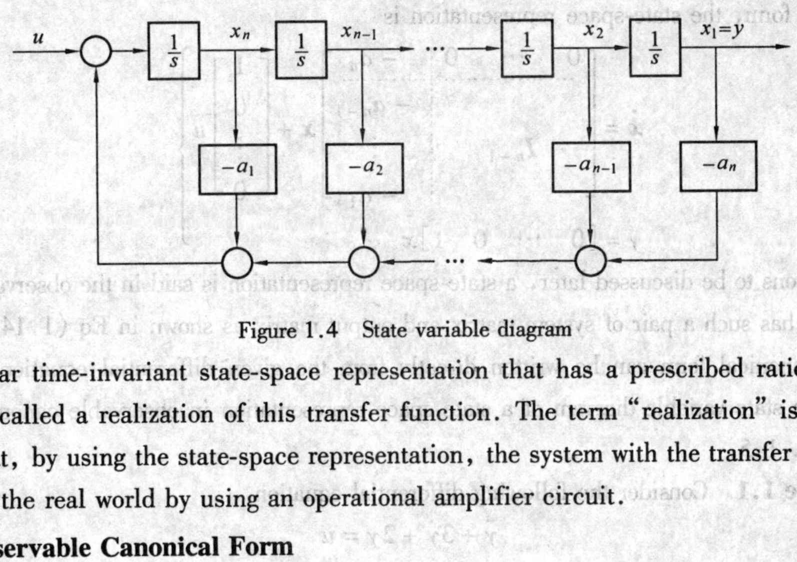

Figure 1.4 State variable diagram

A linear time-invariant state-space representation that has a prescribed rational transfer function is called a realization of this transfer function. The term "realization" is justified by the fact that, by using the state-space representation, the system with the transfer function can be built in the real world by using an operational amplifier circuit.

2. Observable Canonical Form

Taking Laplace transform of Eq. (1.5) under nonzero initial conditions and grouping the terms associated with the same power of s, we get

$$\begin{aligned}(s^n + a_1 s^{n-1} + \cdots + a_n)Y(s) = \\ U(s) + y(0)s^{n-1} + [\dot{y}(0) + a_1 y(0)]s^{n-2} + \cdots + \\ [y^{(n-2)}(0) + a_1 y^{(n-3)}(0) + \cdots + a_{n-2}y(0)]s + \\ [y^{(n-1)}(0) + a_1 y^{(n-2)}(0) + \cdots + a_{n-1}y(0)]\end{aligned}$$

By inspection, based on above equation, the state variables can be chosen as

$$\left.\begin{aligned} x_n &= y \\ x_{n-1} &= \dot{y} + a_1 y \\ &\vdots \\ x_2 &= y^{(n-2)} + a_1 y^{(n-3)} + \cdots + a_{n-2}y \\ x_1 &= y^{(n-1)} + a_1 y^{(n-2)} + \cdots + a_{n-1}y \end{aligned}\right\} \tag{1.11}$$

Thus, from Eq. (1.5) and Eq. (1.11), we get another set of first-order differential equations

$$\left.\begin{aligned} \dot{x}_1 &= u - a_n y = u - a_n x_n \\ \dot{x}_2 &= x_1 - a_{n-1}y = x_1 - a_{n-1}x_n \\ &\vdots \\ \dot{x}_{n-1} &= x_{n-2} - a_2 x_n \\ \dot{x}_n &= x_{n-1} - a_1 x_n \end{aligned}\right\} \tag{1.12}$$

The output equation is

$$y = x_n \tag{1.13}$$

In the matrix form, the state-space representation is

$$\left.\begin{aligned} \dot{\boldsymbol{x}} &= \left[\begin{array}{ccc:c} 0 & \cdots & 0 & -a_n \\ \hdashline & & & -a_{n-1} \\ & \boldsymbol{I}_{n-1} & & \vdots \\ & & & -a_1 \end{array}\right]\boldsymbol{x} + \begin{bmatrix} 1 \\ 0 \\ \vdots \\ 0 \end{bmatrix} u \\ y &= [0 \ \cdots \ 0 \ 1]\boldsymbol{x} \end{aligned}\right\} \tag{1.14}$$

For reasons to be discussed later, a state-space representation is said in the observable canonical form if it has such a pair of system matrix and output matrix as shown in Eq (1.14). Furthermore, this canonical form can be written directly from the given differential equation or transfer function. The state variable diagram of a state-space representation in observable canonical form is shown in Fig. 1.5.

Example 1.1 Consider the following differential equation

$$\ddot{y} + 3\dot{y} + 2y = u$$

Write its state-space representation in controllable canonical form and observable canonical form, respectively.

Solution Defining $x_1 = y$ and $x_2 = \dot{y}$ yields the state-space representation in controllable canonical form

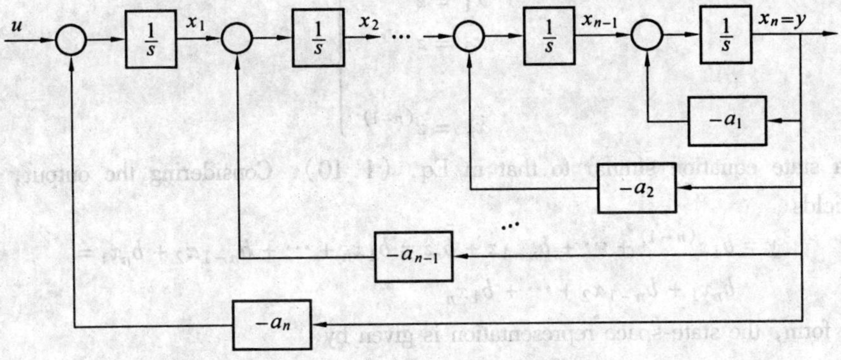

Figure 1.5 State variable diagram for observable canonical form

$$\dot{\boldsymbol{x}} = \begin{bmatrix} 0 & 1 \\ -2 & -3 \end{bmatrix} \boldsymbol{x} + \begin{bmatrix} 0 \\ 1 \end{bmatrix} u$$

$$y = [1 \quad 0] \boldsymbol{x}$$

Similarly, by defining $x_1 = \dot{y} + 3y$ and $x_2 = y$, we have the state-space representation in observable canonical form

$$\dot{\boldsymbol{x}} = \begin{bmatrix} 0 & -2 \\ 1 & -3 \end{bmatrix} \boldsymbol{x} + \begin{bmatrix} 1 \\ 0 \end{bmatrix} u$$

$$y = [0 \quad 1] \boldsymbol{x}$$

1.2.2 Input with Differential Items

Now consider the system described with the differential equation

$$y^{(n)} + a_1 y^{(n-1)} + \cdots + a_{n-1}\dot{y} + a_n y = b_1 u^{(n-1)} + \cdots + b_{n-1}\dot{u} + b_n u \tag{1.15}$$

where $y(t)$ is the output variable and $u(t)$ is the input. The corresponding transfer function of this system is

$$\frac{Y(s)}{U(s)} = \frac{b_1 s^{n-1} + \cdots + b_{n-1}s + b_n}{s^n + a_1 s^{n-1} + \cdots + a_{n-1}s + a_n} \tag{1.16}$$

1. Controllable Canonical Form

Rewrite Eq. (1.16) as

$$\frac{Y(s)}{U(s)} = \frac{Y(s)}{Z(s)} \cdot \frac{Z(s)}{U(s)} \tag{1.17}$$

where

$$\frac{Z(s)}{U(s)} = \frac{1}{s^n + a_1 s^{n-1} + \cdots + a_{n-1}s + a_n} \tag{1.18}$$

$$\frac{Y(s)}{Z(s)} = b_1 s^{n-1} + \cdots + b_{n-1}s + b_n \tag{1.19}$$

From Eq. (1.18) we have

$$z^{(n)} + a_1 z^{(n-1)} + \cdots + a_n z = u \tag{1.20}$$

Choosing state variables

$$\left.\begin{aligned} x_1 &= z \\ x_2 &= \dot{z} \\ &\vdots \\ x_n &= z^{(n-1)} \end{aligned}\right\}$$

results in a state equation similar to that in Eq. (1.10). Considering the output, from Eq. (1.19), yields

$$y = b_1 z^{(n-1)} + \cdots + b_{n-1}\dot{z} + b_n z = b_1 x_n + \cdots + b_{n-1}x_2 + b_n x_1 = \\ b_n x_1 + b_{n-1}x_2 + \cdots + b_1 x_n \tag{1.21}$$

In a matrix form, the state-space representation is given by

$$\left.\begin{aligned} \dot{\boldsymbol{x}} &= \left[\begin{array}{c|ccc} 0 & & & \\ \vdots & & \boldsymbol{I}_{n-1} & \\ 0 & & & \\ \hline -a_n & -a_{n-1} & \cdots & -a_1 \end{array}\right]\boldsymbol{x} + \begin{bmatrix} 0 \\ \vdots \\ 0 \\ 1 \end{bmatrix} u \\ y &= [\,b_n \quad b_{n-1} \quad \cdots \quad b_1\,]\boldsymbol{x} \end{aligned}\right\} \tag{1.22}$$

Obviously, this state-space representation is in the controllable canonical form and also can be written directly from Eq. (1.15) or Eq. (1.16).

2. Observable Canonical Form

In a similar way as that in section 1.2.1, taking Laplace transform of Eq. (1.15) under nonzero initial conditions and grouping the terms associated with the same power of s yield

$$\begin{aligned} &(s^n + a_1 s^{n-1} + \cdots + a_n)Y(s) = \\ &(b_1 s^{n-1} + b_2 s^{n-2} + \cdots + b_n)U(s) + y(0)s^{n-1} + [\dot{y}(0) + a_1 y(0) - b_1 u(0)]s^{n-2} + \cdots + \\ &[y^{(n-2)}(0) + a_1 y^{(n-3)}(0) + \cdots + a_{n-2}y(0) - b_1 u^{(n-3)}(0) - \cdots - b_{n-2}u(0)]s + \\ &[y^{(n-1)}(0) + a_1 y^{(n-2)}(0) + \cdots + a_{n-1}y(0) - b_1 u^{(n-2)}(0) - \cdots - b_{n-1}u(0)] \end{aligned} \tag{1.23}$$

Choosing the state variables

$$\left.\begin{aligned} x_n &= y \\ x_{n-1} &= \dot{y} + a_1 y - b_1 u \\ &\vdots \\ x_2 &= y^{(n-2)} + a_1 y^{(n-3)} + \cdots + a_{n-2}y - b_1 u^{(n-3)} - \cdots - b_{n-2}u \\ x_1 &= y^{(n-1)} + a_1 y^{(n-2)} + \cdots + a_{n-1}y - b_1 u^{(n-2)} - \cdots - b_{n-1}u \end{aligned}\right\} \tag{1.24}$$

results in a state-space representation in observable canonical form

$$\left.\begin{aligned} \dot{\boldsymbol{x}} &= \left[\begin{array}{ccc|c} 0 & \cdots & 0 & -a_n \\ \hline & & & -a_{n-1} \\ & \boldsymbol{I}_{n-1} & & \vdots \\ & & & -a_1 \end{array}\right]\boldsymbol{x} + \begin{bmatrix} b_n \\ b_{n-1} \\ \vdots \\ b_1 \end{bmatrix} u \\ y &= [0 \quad \cdots \quad 0 \quad 1]\boldsymbol{x} \end{aligned}\right\} \tag{1.25}$$

Example 1.2 Consider the transfer function

$$\frac{Y(s)}{U(s)} = \frac{11s + 6}{s^3 + 6s^2 + 11s + 6}$$

Write its state-space representation in controllable canonical form and observable canonical form, respectively.

Solution Letting $\dddot{z} + 6\ddot{z} + 11\dot{z} + 6z = u$ and choosing state variables as

$$x_1 = z$$

$$x_2 = \dot{z}$$

$$x_3 = \ddot{z}$$

we obtain a state-space representation in controllable canonical form

$$\dot{\boldsymbol{x}} = \begin{bmatrix} 0 & 1 & 0 \\ 0 & 0 & 1 \\ -6 & -11 & -6 \end{bmatrix} \boldsymbol{x} + \begin{bmatrix} 0 \\ 0 \\ 1 \end{bmatrix} u$$

$$y = [6 \quad 11 \quad 0] \boldsymbol{x}$$

Choosing state variables as

$$x_3 = y$$

$$x_2 = \dot{y} + 6y$$

$$x_1 = \ddot{y} + 6\dot{y} + 11y - 11u$$

yields the state-space representation in observable canonical form

$$\dot{\boldsymbol{x}} = \begin{bmatrix} 0 & 0 & -6 \\ 1 & 0 & -11 \\ 0 & 1 & -6 \end{bmatrix} \boldsymbol{x} + \begin{bmatrix} 6 \\ 11 \\ 0 \end{bmatrix} u$$

$$y = [0 \quad 0 \quad 1] \boldsymbol{x}$$

3. Jordan Canonical Form

It is particular useful if the state equation may be written such that the system matrix takes the Jordan form. Correspondingly, the state-space representation is of the Jordan canonical form.

Case 1 All the poles of the transfer function are distinct.

At first we consider the simple case that all the poles, $s_1, s_2, \cdots, s_n$, of the transfer function given by Eq. (1.16) are distinct. Then, the transfer function can be expanded by partial fraction expansion into

$$\frac{Y(s)}{U(s)} = \frac{M(s)}{(s - s_1)(s - s_2)\cdots(s - s_n)} = \sum_{i=1}^{n} \frac{c_i}{s - s_i}$$

where

$$c_i = \lim_{s \to s_i} (s - s_i) \frac{Y(s)}{U(s)}$$

Choose the state variable so that

$$X_i(s) = \frac{1}{s - s_i} U(s) \qquad i = 1, 2, \cdots, n \tag{1.26}$$

which are called the typical functions. Thus we have

$$sX_i(s) = s_iX_i(s) + U(s)$$

or

$$\dot{x}(t) = s_ix_i(t) + u(t)$$

and the output is

$$y(t) = \sum_{i=1}^{n} c_ix_i(t)$$

In this way, the state-space representation is written as

$$\left.\begin{aligned}\dot{\boldsymbol{x}} &= \begin{bmatrix} s_1 & 0 & \cdots & 0 \\ 0 & s_2 & \cdots & 0 \\ \vdots & \vdots & & \vdots \\ 0 & 0 & \cdots & s_n \end{bmatrix}\boldsymbol{x} + \begin{bmatrix} 1 \\ 1 \\ \vdots \\ 1 \end{bmatrix}u \\ y &= \begin{bmatrix} c_1 & c_2 & \cdots & c_n \end{bmatrix}\boldsymbol{x}\end{aligned}\right\} \tag{1.27}$$

Note that in the system matrix the leading diagonal elements are just the system poles and only the leading diagonal elements may be nonzero. In this case, since all of the poles are distinct, the system matrix becomes a diagonal matrix ($\boldsymbol{\Lambda}$ matrix).

The state variable diagram for Eq. (1.27) is shown in Fig. 1.6.

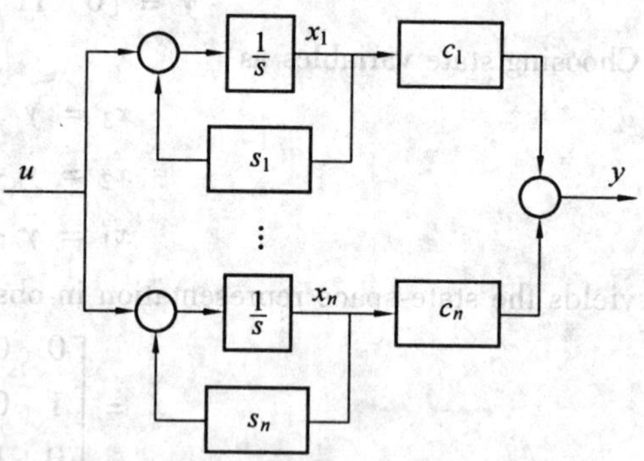

Figure 1.6 State-variable diagram for Eq. (1.27)

Example 1.3 Again consider the transfer function

$$\frac{Y(s)}{U(s)} = \frac{11s + 6}{s^3 + 6s^2 + 11s + 6}$$

given in Example 1.2. Write its state-space representation in the Jordan canonical form.

Solution The transfer function can be expanded by partial fraction expansion into

$$\frac{Y(s)}{U(s)} = \frac{11s + 6}{(s + 1)(s + 2)(s + 3)} = -\frac{5}{2}\frac{1}{s + 1} + 16\frac{1}{s + 2} - \frac{17}{2}\frac{1}{s + 3}$$

The state-space representation in Jordan canonical form is

$$\dot{\boldsymbol{x}} = \begin{bmatrix} -1 & 0 & 0 \\ 0 & -2 & 0 \\ 0 & 0 & -3 \end{bmatrix}\boldsymbol{x} + \begin{bmatrix} 1 \\ 1 \\ 1 \end{bmatrix}u$$

$$y = \begin{bmatrix} -\frac{5}{2} & 16 & -\frac{17}{2} \end{bmatrix}\boldsymbol{x}$$

where the state variables satisfy

$$X_1(s) = \frac{1}{s + 1}U(s),\ X_2(s) = \frac{1}{s + 2}U(s),\ X_3(s) = \frac{1}{s + 3}U(s)$$

Case 2 The poles of transfer function are not all distinct.

Unlike the previous case, if a transfer function has repeated poles, it is not always possible to establish a state-space representation with the diagonal system matrix. We will use examples to illustrate the procedure to derive the Jordan canonical form. The idea can be easily extended to the general cases.

For convenience, we assume that the poles of transfer function are s_1 with multiplicity r and all other poles are distinct. Then, the transfer function can be expanded by partial fraction expansion into

$$\frac{Y(s)}{U(s)} = \sum_{i=1}^{r} \frac{c_{1i}}{(s - s_1)^i} + \sum_{j=r+1}^{n} \frac{c_j}{s - s_j}$$

where

$$c_{1i} = \frac{1}{(r-i)!} \lim_{s \to s_1} \frac{\mathrm{d}^{r-i}}{\mathrm{d}s^{r-i}} \left[(s - s_1)^r \frac{Y(s)}{U(s)} \right], \quad i = 1,2,\cdots,r$$

$$c_j = \lim_{s \to s_j} (s - s_j) \frac{Y(s)}{U(s)}, \quad j = r+1, r+2, \cdots, n$$

Choose the typical functions as

$$X_i(s) = \frac{1}{(s - s_1)^{r-i+1}} U(s), \quad i = 1,2,\cdots,r \tag{1.28}$$

$$X_j(s) = \frac{1}{s - s_j} U(s), \quad j = r+1, r+2, \cdots, n \tag{1.29}$$

and then consider the expressions for $\dot{x}_k(t)$, $k = 1,2,\cdots,r-1,r,r+1,\cdots,n$

(1) $k < r$

$$X_k(s) = \frac{1}{(s - s_1)^{r-k+1}} U(s) = \frac{1}{(s - s_1)(s - s_1)^{r-(k+1)+1}} U(s) = \frac{1}{s - s_1} X_{k+1}(s)$$

$$sX_k(s) = s_1 X_k(s) + X_{k+1}(s)$$

$$\dot{x}_k = s_1 x_k + x_{k+1}$$

(2) $k = r$

$$X_k(s) = \frac{1}{s - s_1} U(s)$$

$$\dot{x}_k = s_1 x_k + u$$

(3) $k > r$

$$X_k(s) = \frac{1}{s - s_k} U(s)$$

$$\dot{x}_k = s_k x_k + u$$

Writing above first-order differential equations into the matrix form yields the state equation

$$
\begin{bmatrix} \dot{x}_1 \\ \dot{x}_2 \\ \vdots \\ \dot{x}_{r-1} \\ \dot{x}_r \\ \dot{x}_{r+1} \\ \vdots \\ \dot{x}_n \end{bmatrix} =
\left[\begin{array}{ccccc|ccc}
s_1 & 1 & \cdots & 0 & 0 & 0 & \cdots & 0 \\
0 & s_1 & \cdots & 0 & 0 & 0 & \cdots & 0 \\
\vdots & \vdots & & \vdots & \vdots & \vdots & & \vdots \\
0 & 0 & \cdots & s_1 & 1 & 0 & \cdots & 0 \\
0 & 0 & \cdots & 0 & s_1 & 0 & \cdots & 0 \\
\hline
0 & 0 & \cdots & 0 & 0 & s_{r+1} & \cdots & 0 \\
\vdots & \vdots & & \vdots & \vdots & \vdots & & \vdots \\
0 & 0 & \cdots & 0 & 0 & 0 & \cdots & s_n
\end{array}\right]
\begin{bmatrix} x_1 \\ x_2 \\ \vdots \\ x_{r-1} \\ x_r \\ x_{r+1} \\ \vdots \\ x_n \end{bmatrix} +
\begin{bmatrix} 0 \\ 0 \\ \vdots \\ 0 \\ 1 \\ 1 \\ \vdots \\ 1 \end{bmatrix} u \qquad (1.30)
$$

As for the output, we have

$$
Y(s) = \sum_{i=1}^{r} \frac{c_{1i}}{(s-s_1)^i} U(s) + \sum_{j=r+1}^{n} \frac{c_j}{s-s_j} U(s) =
$$

$$
c_{11}X_r(s) + c_{12}X_{r-1}(s) + \cdots + c_{1r}X_1(s) + c_{r+1}X_{r+1}(s) + \cdots + c_nX_n(s)
$$

$$
y(t) = c_{1r}x_1(t) + \cdots + c_{11}x_r(t) + c_{r+1}x_{r+1}(t) + \cdots + c_nx_n(t)
$$

i.e. the output equation in the matrix form is

$$
y = \left[c_{1r} \quad c_{1,r-1} \quad \cdots \quad c_{12} \quad c_{11} \;\vdots\; c_{r+1} \quad \cdots \quad c_n \right] \boldsymbol{x} \qquad (1.31)
$$

Note that the leading diagonal of the system matrix consists of the Jordan blocks and the remaining elements are zeros. A diagonal system matrix is a special case of the Jordan form: all of its Jordan blocks are of order 1. The state variable diagram for this case is shown in Fig. 1.7.

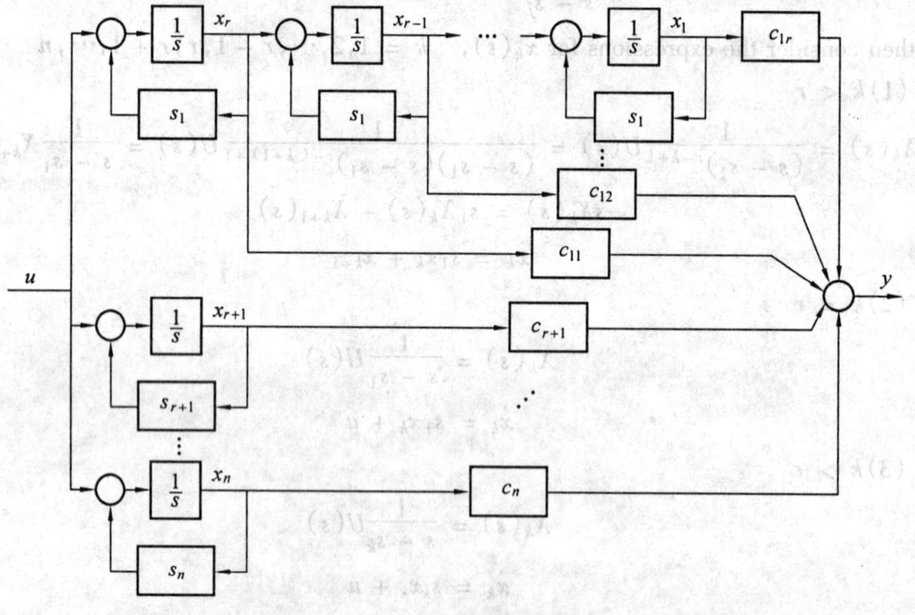

Figure 1.7 State variable diagram for Eq. (1.30) and Eq. (1.31)

Since the leading diagonal elements are just the system poles, i.e. the eigenvalues of the system matrix, the Jordan canonical form is sometimes called the eigenvalue canonical form.

Example 1.4 Find the Jordan canonical form of state-space representation if the system is described by a transfer function

$$\frac{Y(s)}{U(s)} = \frac{2s^2 + 6s + 5}{s^3 + 4s^2 + 5s + 2}$$

Solution This transfer function can be expanded by partial fraction expansion into

$$\frac{Y(s)}{U(s)} = \frac{2s^2 + 6s + 5}{(s + 1)^2(s + 2)} = \frac{c_{11}}{(s + 1)^2} + \frac{c_{12}}{s + 1} + \frac{c_3}{s + 2}$$

where

$$c_{12} = \lim_{s \to -1} \frac{2s^2 + 6s + 5}{s + 2} = 1, \quad c_{11} = \lim_{s \to -1} \frac{d}{ds} \frac{2s^2 + 6s + 5}{s + 2} = 1$$

$$c_3 = \lim_{s \to -2} \frac{2s^2 + 6s + 5}{(s + 1)^2} = 1$$

i.e.

$$\frac{Y(s)}{U(s)} = \frac{1}{(s + 1)^2} + \frac{1}{s + 1} + \frac{1}{s + 2}$$

From Eq. (1.30) we have

$$\dot{\boldsymbol{x}} = \left[\begin{array}{cc:c} -1 & 1 & 0 \\ \hdashline 0 & -1 & 0 \\ 0 & 0 & -2 \end{array}\right] \boldsymbol{x} + \left[\begin{array}{c} 0 \\ 1 \\ \hdashline 1 \end{array}\right] u$$

$$y = [1 \quad 1 \quad 1] \boldsymbol{x}$$

where the state variables satisfy

$$X_1(s) = \frac{1}{(s + 1)^2} U(s), \quad X_2(s) = \frac{1}{s + 1} U(s), \quad X_3(s) = \frac{1}{s + 2} U(s)$$

1.3 State-Space Description from Block Diagram

Sometimes, we establish the state-space representation directly from the given block diagrams so that the outputs of the blocks may be chosen as the state variables to reflect the actual physical variables. This procedure is illustrated with an example.

Example 1.5 Write the state-space representation for the system shown in Fig. 1.8(a).

Solution At first we redraw the block diagram, as shown in Fig. 1.8(b), so that each block can be simulated with one integrator and/or multiplicator.

Choosing the outputs of the blocks containing first-order elements as the state variables as shown in Fig. 1.8(b), we get

$$\begin{bmatrix} \dot{x}_1 \\ \dot{x}_2 \\ \dot{x}_3 \\ x_4 \end{bmatrix} = \begin{bmatrix} -2 & 0 & -20 & 0 \\ 1 & 0 & 0 & -1 \\ 0 & 2 & -1 & 0 \\ 0 & 2 & -1 & -3 \end{bmatrix} \begin{bmatrix} x_1 \\ x_2 \\ x_3 \\ x_4 \end{bmatrix} + \begin{bmatrix} 10 \\ 0 \\ 0 \\ 0 \end{bmatrix} u$$

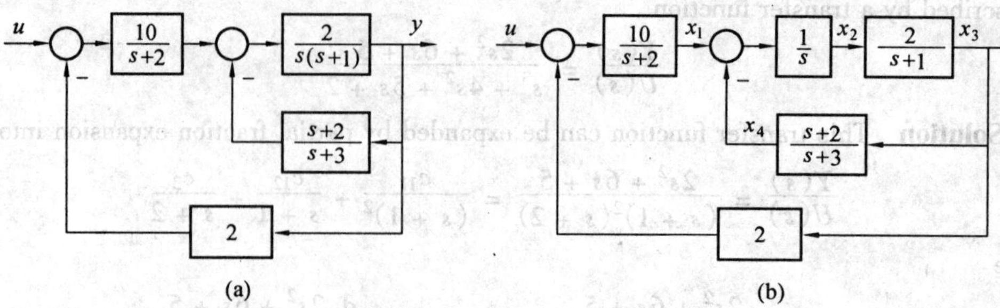

Figure 1.8 Block diagram for Example 1.5

$$y = \begin{bmatrix} 0 & 0 & 1 & 0 \end{bmatrix} \begin{bmatrix} x_1 \\ x_2 \\ x_3 \\ x_4 \end{bmatrix}$$

1.4 Transfer Function from State-Space Representation

A procedure was given above for writing the state-space representation of a system if the transfer function, or differential equation, is known. In some cases, the transfer function (or the transfer function matrix for a multiple-variable system) will need to be evaluated from a state-space representation. The problem may be stated as: given coefficient matrixes $\boldsymbol{A}$, $\boldsymbol{B}$, $\boldsymbol{C}$, and $\boldsymbol{D}$, find the relationship between output y and input u.

We begin with the typical form of the state-space representation for a single-variable system

$$\left.\begin{aligned} \dot{\boldsymbol{x}}(t) &= \boldsymbol{A}\boldsymbol{x}(t) + \boldsymbol{B}u(t) \\ y(t) &= \boldsymbol{C}\boldsymbol{x}(t) \end{aligned}\right\} \tag{1.32}$$

Taking Laplace transform of the state equation in Eq. (1.32), letting the initial conditions be zero, we have

$$s\boldsymbol{X}(s) = \boldsymbol{A}\boldsymbol{X}(s) + \boldsymbol{B}U(s) \tag{1.33}$$

This equation may be written as

$$(s\boldsymbol{I} - \boldsymbol{A})\boldsymbol{X}(s) = \boldsymbol{B}U(s)$$

and is easily solved for $\boldsymbol{X}(s)$

$$\boldsymbol{X}(s) = (s\boldsymbol{I} - \boldsymbol{A})^{-1}\boldsymbol{B}U(s) \tag{1.34}$$

Substituting above expression for $\boldsymbol{X}(s)$ into the output equation in Eq. (1.32) yields

$$Y(s) = \boldsymbol{C}(s\boldsymbol{I} - \boldsymbol{A})^{-1}\boldsymbol{B}U(s) \tag{1.35}$$

Hence, the transfer function of the system is given by

$$G(s) = \frac{Y(s)}{U(s)} = \boldsymbol{C}(s\boldsymbol{I} - \boldsymbol{A})^{-1}\boldsymbol{B} \tag{1.36}$$

Example 1.6 Find the transfer function for the system described by the following state-space representation

$$\dot{x} = \begin{bmatrix} 0 & 1 & 0 \\ 0 & -1 & -1 \\ 0 & 0 & -3 \end{bmatrix} x + \begin{bmatrix} 0 \\ 1 \\ 1 \end{bmatrix} u$$

$$y = [1 \quad 0 \quad 0] x$$

Solution Since

$$\det(sI - A) = \begin{vmatrix} s & -1 & 0 \\ 0 & s+1 & 1 \\ 0 & 0 & s+3 \end{vmatrix} = s(s+1)(s+3)$$

$$\operatorname{adj}(sI - A) = \begin{bmatrix} (s+1)(s+3) & s+3 & -1 \\ 0 & s(s+3) & -s \\ 0 & 0 & s(s+1) \end{bmatrix}$$

we obtain

$$G(s) = \frac{Y(s)}{U(s)} = C(sI - A)^{-1}B = C\frac{\operatorname{adj}(sI - A)}{\det(sI - A)}B =$$

$$[1 \quad 0 \quad 0]\frac{1}{s(s+1)(s+3)}\begin{bmatrix} (s+1)(s+3) & s+3 & -1 \\ 0 & s(s+3) & -s \\ 0 & 0 & s(s+1) \end{bmatrix}\begin{bmatrix} 0 \\ 1 \\ 1 \end{bmatrix} =$$

$$\frac{s+2}{s(s+1)(s+3)}$$

For the multivariable case, the input $U(s)$ is a $p \times 1$ vector and the output $Y(s)$ is a $q \times 1$ vector. The preceding derivation remains valid, and Eq. (1.35) becomes

$$Y(s) = C(sI - A)^{-1}BU(s) = G(s)U(s)$$

that is

$$G(s) = C(sI - A)^{-1}B \tag{1.37}$$

where $G(s)$ is now a $q \times p$ matrix. The element of $G(s)$, $G_{ij}(s)$, is the transfer function,

$$G_{ij}(s) = \frac{Y_i(s)}{U_j(s)} \tag{1.38}$$

with all other inputs equal to zero.

1.5 Characteristic Polynomial and Eigenvalues

The selection of state variables for a particular system is not unique. Hence, the state-space representation of a system is not unique.

At first, we consider the effect of linear nonsingular transformation on the state-space representation. Any new state variables may be defined as a simple linear transformation of the old state variables. Assume that a system is described by the state-space representation

$$\left.\begin{aligned} \dot{x}(t) &= Ax(t) + Bu(t) \\ y(t) &= Cx(t) \end{aligned}\right\} \tag{1.39}$$

we may construct a new set of state variables z from the old ones x by

$$x = Pz \tag{1.40}$$

where P is a $n \times n$ nonsingular matrix. Substituting for z into the state equation gets

$$P\dot{z} = APz + Bu$$

or

$$\dot{z} = P^{-1}APz + P^{-1}Bu \tag{1.41}$$

Similarly, substituting for z into the output equation gets

$$y = CPz \tag{1.42}$$

Thus, we can rewrite the state and output equations in terms of the new state variables as

$$\left.\begin{aligned} \dot{z} &= \bar{A}z + \bar{B}u \\ y &= \bar{C}z \end{aligned}\right\} \tag{1.43}$$

where

$$\bar{A} = P^{-1}AP, \quad \bar{B} = P^{-1}B, \quad \bar{C} = CP$$

For a given state-space representation, we can write Eq. (1.37) as

$$G(s) = C\frac{\operatorname{adj}(sI - A)}{\det(sI - A)}B \tag{1.44}$$

where the denominator of $G(s)$, $\det(sI - A)$ is called the characteristic polynomial of the system matrix.

We will now show that the characteristic polynomial of a system matrix is unchanged by the linear nonsingular transformation. For the characteristic polynomial of the new system matrix, we have

$$\begin{aligned} \det(sI - \bar{A}) &= \det(sI - P^{-1}AP) = \det(sP^{-1}P - P^{-1}AP) = \\ &\det[P^{-1}(sI - A)P] = \det(P^{-1}P) \cdot \det(sI - A) = \\ &\det(sI - A) \end{aligned}$$

This result confirms that the characteristic polynomial of a system matrix is independent of the state variables chosen to represent the system. In fact, it can be seen from Eq. (1.44) that the characteristic polynomial of a system matrix just is the characteristic polynomial of $G(s)$.

The roots of a characteristic polynomial, which are the poles of a transfer function, are called the eigenvalues of the system matrix, or the system eigenvalues, in the state-space description. Thus, we can get an important conclusion that the system eigenvalues are independent of the chosen state variables.

Problems

P1.1 The armature-controlled dc motor can be described by the following equations

$$v_a = L_a\frac{di_a}{dt} + R_a i_a + e_b$$

$$e_b = C_e\frac{d\theta_m}{dt}$$

$$T_m = C_m i_a$$

$$T_m = J_m\frac{d^2\theta_m}{dt^2} + f_m\frac{d\theta_m}{dt}$$

or by the transfer function

$$\frac{\theta_m(s)}{V_a(s)} = \frac{C_m}{s[L_a J_m s^2 + (L_a f_m + R_a J_m)s + (R_a f_m + C_e C_m)]}$$

(a) Let the state variable be $x_1 = \theta_m$, $x_2 = \dot{\theta}_m$, $x_3 = \ddot{\theta}_m$, and the output be $y = \theta_m$. Write the state equation in matrix form.

(b) Let the state variable be $\bar{x}_1 = i_a$, $\bar{x}_2 = \theta_m$, $\bar{x}_3 = \dot{\theta}_m$, and $y = \theta_m$. Write the state equation in matrix form.

(c) Determine the transformation matrix $\boldsymbol{T}$ so that $\boldsymbol{x} = \boldsymbol{T}\bar{\boldsymbol{x}}$.

P1.2 A circuit with two inputs, one voltage source and one current source, is shown in Fig. P1.2. Let the state variable be $x_1 = i_L$, $x_2 = v_{C_1}$, $x_3 = v_{C_2}$, and the output be $y = \theta_m$. Determine the state-space representation for the circuit shown.

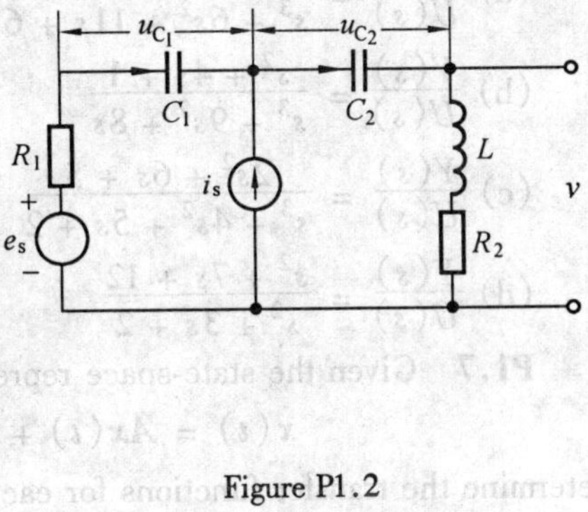

Figure P1.2

P1.3 The block diagrams of three systems are shown in Fig. P1.3, where x_1, x_2, and x_3 are specified state variables. Determine the state-space representations for each system.

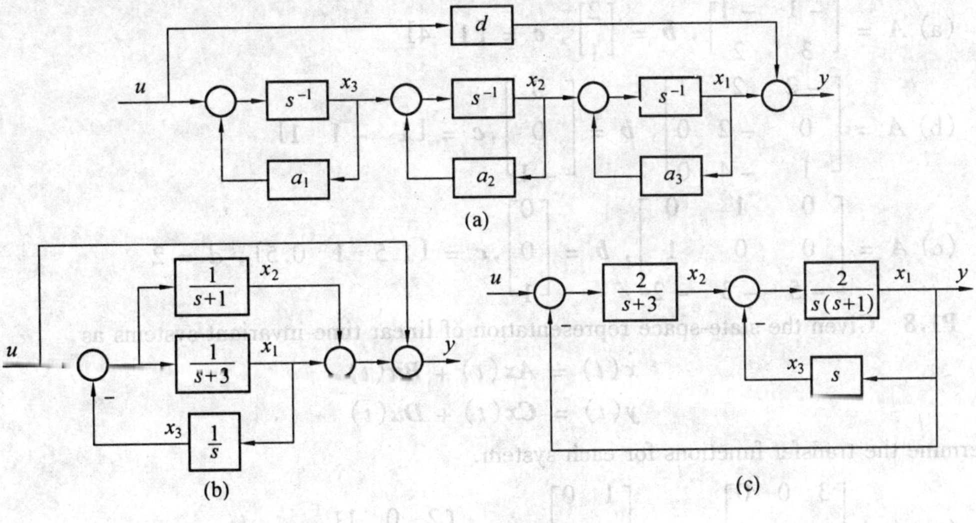

Figure P1.3

P1.4 Obtain the state-space representations for the following systems

(a) $\dddot{y} + 3\ddot{y} + 2\dot{y} + 2y = u$

(b) $\dddot{y} - 3\ddot{y} + 4\dot{y} - 2y = 2u$

(c) $\dddot{y} + 3\ddot{y} + 2\dot{y} + 2y = 0$

(d) $\dddot{y} - 3a\ddot{y} + 3a^2\dot{y} - a^3 y = u$

(e) $\begin{cases} \ddot{y}_1 + 4\dot{y}_1 - 3y_2 = u_1 \\ \dot{y}_2 + \dot{y}_1 + y_1 + 2y_2 = u_2 \end{cases}$

P1.5 Obtain the state-space representations for the following systems

(a) $2\ddot{y} - 3y = \ddot{u} - 2u$

(b) $\dddot{y} + 4\ddot{y} + 2\dot{y} + y = \ddot{u} + \dot{u} + 3u$

(c) $\dddot{y} + 10\ddot{y} + 80\dot{y} + 128y = 5\dot{u} + 6u$

(d) $\dddot{y} + 6\ddot{y} + 11\dot{y} + 6y = \dddot{u} + 8\ddot{u} + 17\dot{u} + u$

P1.6 Determine the state-space representations for the systems described by the following transfer functions.

(a) $\dfrac{Y(s)}{U(s)} = \dfrac{2s^2 + 18s + 40}{s^3 + 6s^2 + 11s + 6}$

(b) $\dfrac{Y(s)}{U(s)} = \dfrac{s^2 + 4s + 1}{s^3 + 9s^2 + 8s}$

(c) $\dfrac{Y(s)}{U(s)} = \dfrac{2s^2 + 6s + 5}{s^3 + 4s^2 + 5s + 2}$

(d) $\dfrac{Y(s)}{U(s)} = \dfrac{s^2 + 7s + 12}{s^2 + 3s + 2}$

P1.7 Given the state-space representation of linear time-invariant systems as

$$\dot{\boldsymbol{x}}(t) = \boldsymbol{A}\boldsymbol{x}(t) + \boldsymbol{b}u(t), \quad y(t) = \boldsymbol{c}\boldsymbol{x}(t) + du(t)$$

determine the transfer functions for each system.

(a) $\boldsymbol{A} = \begin{bmatrix} -1 & -1 \\ 3 & 2 \end{bmatrix}$, $\boldsymbol{b} = \begin{bmatrix} 2 \\ 1 \end{bmatrix}$, $\boldsymbol{c} = [1 \quad 4]$

(b) $\boldsymbol{A} = \begin{bmatrix} -2 & 2 & 1 \\ 0 & -2 & 0 \\ 1 & -4 & 0 \end{bmatrix}$, $\boldsymbol{b} = \begin{bmatrix} 0 \\ 0 \\ -1 \end{bmatrix}$, $\boldsymbol{c} = [1 \quad -1 \quad 1]$

(c) $\boldsymbol{A} = \begin{bmatrix} 0 & 1 & 0 \\ 0 & 0 & 1 \\ -5 & -3 & -2 \end{bmatrix}$, $\boldsymbol{b} = \begin{bmatrix} 0 \\ 0 \\ 1 \end{bmatrix}$, $\boldsymbol{c} = [1.5 \quad 1 \quad 0.5]$, $d = 2$

P1.8 Given the state-space representation of linear time-invariant systems as

$$\dot{\boldsymbol{x}}(t) = \boldsymbol{A}\boldsymbol{x}(t) + \boldsymbol{B}\boldsymbol{u}(t)$$
$$\boldsymbol{y}(t) = \boldsymbol{C}\boldsymbol{x}(t) + \boldsymbol{D}\boldsymbol{u}(t)$$

determine the transfer functions for each system.

(a) $\boldsymbol{A} = \begin{bmatrix} 3 & 0 & 0 \\ 1 & 5 & 2 \\ 0 & 2 & 1 \end{bmatrix}$, $\boldsymbol{B} = \begin{bmatrix} 1 & 0 \\ 2 & 0 \\ 0 & 5 \end{bmatrix}$, $\boldsymbol{C} = \begin{bmatrix} 2 & 0 & 1 \\ 6 & 2 & 0 \end{bmatrix}$

(b) $\boldsymbol{A} = \begin{bmatrix} 0 & 1 & 0 \\ 0 & 0 & 1 \\ -1 & -1 & -2 \end{bmatrix}$, $\boldsymbol{B} = \begin{bmatrix} 0 & 0 \\ 1 & 0 \\ 0 & 1 \end{bmatrix}$, $\boldsymbol{C} = \begin{bmatrix} 1 & 0 & 0 \\ 0 & 0 & 1 \end{bmatrix}$, $\boldsymbol{D} = \begin{bmatrix} 0 & 1 \\ 1 & 1 \end{bmatrix}$

Chapter 2 Dynamic Analysis of Linear Systems

In the preceding chapter we introduced the state-space description of control systems. Obviously, the next step in system analysis is to find the solution of the state equation.

2.1 Time-Invariant Continuous Systems

2.1.1 Solution to Homogeneous Equation

At first, let's consider the linear time-invariant continuous systems. Since the solution of the homogeneous part of a differential equation essentially also determines the solution of inhomogeneous part of a differential equation, we shall start with the homogeneous equation

$$\dot{x}(t) = Ax(t),\ t \geqslant 0;\ x(0) = x_0 \tag{2.1}$$

Here, as usually in the time-invariant systems, it is assumed that the initial time is $t_0 = 0$.

1. Infinite Series Solution

One method of solving a differential equation is to assume that the solution is an infinite series with unknown coefficients. The infinite series is then substituted into the differential equation to evaluate the unknown coefficients. This method is now used to find the solution of the state equation.

In this way, the solution must be assumed to be of the form

$$x(t) = b_0 + b_1 t + b_2 t^2 + \cdots + b_k t^k + \cdots = \sum_{k=0}^{\infty} b_k t^k,\ t \geqslant 0 \tag{2.2}$$

Substituting Eq. (2.2) into the homogeneous equation yields

$$b_1 + 2b_2 t + 3b_3 t^2 + \cdots + k b_k t^{k-1} + \cdots = A(b_0 + b_1 t + b_2 t^2 + \cdots + b_k t^k + \cdots) \tag{2.3}$$

Equating the coefficients of t^i, $i = 1,2,\cdots$, in Eq. (2.3) results in

$$b_1 = Ab_0,\ b_2 = \frac{1}{2}Ab_1 = \frac{1}{2}A^2 b_0,\ b_3 = \frac{1}{3}Ab_2 = \frac{1}{3!}A^3 b_0, \cdots, b_k = \frac{1}{k!}A^k b_0, \cdots$$

Hence, the homogenous solution can be expressed as

$$x(t) = b_0 + Ab_0 t + \frac{1}{2!}A^2 b_0 t^2 + \cdots + \frac{1}{k!}A^k b_0 t^k + \cdots =$$

$$\left(I_n + At + \frac{1}{2!}A^2 t^2 + \cdots + \frac{1}{k!}A^k t^k + \cdots\right) b_0 \tag{2.4}$$

From Eq. (2.4) and the given initial condition, we get

$$b_0 = x(t)\big|_{t=0} = x_0$$

Therefore, the homogeneous solution is given by

$$\boldsymbol{x}(t) = \left(\boldsymbol{I}_n + \boldsymbol{A}t + \frac{1}{2!}\boldsymbol{A}^2 t^2 + \cdots + \frac{1}{k!}\boldsymbol{A}^k t^k + \cdots \right) \boldsymbol{x}_0 \tag{2.5}$$

2. State Transition Matrix and Matrix Exponential Function

Noting that each term in the parentheses in Eq. (2.5) is a $n \times n$ matrix, we represent the homogeneous solution as

$$\dot{\boldsymbol{x}}(t) = \boldsymbol{\Phi}(t)\boldsymbol{x}_0 \tag{2.6}$$

where the $n \times n$ matrix $\boldsymbol{\Phi}(t)$ is called the state transition matrix, since it completely defines the transition of the state from the initial time $t = 0$ to any time t.

Furthermore, because of the similarity between the infinite series in Eq. (2.5) and the Taylor's series expansion of the scalar exponential, we define the matrix exponential function, or matrix exponential for short, as

$$\mathrm{e}^{\boldsymbol{A}t} = \boldsymbol{I}_n + \boldsymbol{A}t + \frac{1}{2!}\boldsymbol{A}^2 t^2 + \cdots + \frac{1}{k!}\boldsymbol{A}^k t^k + \cdots = \sum_{k=0}^{\infty} \frac{1}{k!}\boldsymbol{A}^k t^k \tag{2.7}$$

Thus, the homogeneous solution for a linear time-invariant state equation can be represented as

$$\boldsymbol{x}(t) = \mathrm{e}^{\boldsymbol{A}t}\boldsymbol{x}_0 \tag{2.8}$$

and, in this case we have

$$\boldsymbol{\Phi}(t) = \mathrm{e}^{\boldsymbol{A}t} \tag{2.9}$$

It should be pointed out that the conclusion given by Eq. (2.9) is only for the linear time-invariant system. The state transition matrix can be used to relate $\boldsymbol{x}(t)$ to its initial condition $\boldsymbol{x}_0$ even for the time-varying systems, but the matrix exponential function is only effective for the time-invariant systems.

3. Operational Properties of $\boldsymbol{\Phi}(t)$

We see that the state transition matrix $\boldsymbol{\Phi}(t)$, or the matrix exponential function $\mathrm{e}^{\boldsymbol{A}t}$, is central to the solution of the linear time-invariant state equation. Since $\boldsymbol{\Phi}(t)$, or $\mathrm{e}^{\boldsymbol{A}t}$, contains the complete information describing the system dynamics, it is useful to be familiar with the following essential operational properties:

(1) $\boldsymbol{\Phi}(0) = \boldsymbol{I}$ (2.10)

Proof $\boldsymbol{\Phi}(0) = \mathrm{e}^{\boldsymbol{A}t}\big|_{t=0} = \boldsymbol{I}$

(2) $\dot{\boldsymbol{\Phi}}(t) = \boldsymbol{A}\boldsymbol{\Phi}(t) = \boldsymbol{\Phi}(t)\boldsymbol{A}$ (2.11)

Proof $\dot{\boldsymbol{\Phi}}(t) = \dfrac{\mathrm{d}}{\mathrm{d}t}\mathrm{e}^{\boldsymbol{A}t} = \boldsymbol{A} + \boldsymbol{A}^2 t + \dfrac{1}{2!}\boldsymbol{A}^2 t^2 + \cdots + \dfrac{1}{(k-1)!}\boldsymbol{A}^k t^{k-1} + \cdots =$

$$\boldsymbol{A} \cdot \mathrm{e}^{\boldsymbol{A}t} = \mathrm{e}^{\boldsymbol{A}t} \cdot \boldsymbol{A} = \boldsymbol{A}\boldsymbol{\Phi}(t) = \boldsymbol{\Phi}(t)\boldsymbol{A}$$

(3) $\boldsymbol{\Phi}(t \pm \tau) = \boldsymbol{\Phi}(t)\boldsymbol{\Phi}(\pm \tau) = \boldsymbol{\Phi}(\pm \tau)\boldsymbol{\Phi}(t)$ (2.12)

Proof $\boldsymbol{\Phi}(t \pm \tau) = \mathrm{e}^{\boldsymbol{A}(t \pm \tau)} = \mathrm{e}^{\boldsymbol{A}t} \cdot \mathrm{e}^{\pm \boldsymbol{A}\tau} = \mathrm{e}^{\pm \boldsymbol{A}\tau} \cdot \mathrm{e}^{\boldsymbol{A}t} =$

$$\boldsymbol{\Phi}(t)\boldsymbol{\Phi}(\pm \tau) = \boldsymbol{\Phi}(\pm \tau)\boldsymbol{\Phi}(t)$$

(4) $\boldsymbol{\Phi}^{-1}(t) = \boldsymbol{\Phi}(-t)$ (2.13)

Proof $\boldsymbol{\Phi}^{-1}(t) = (\mathrm{e}^{\boldsymbol{A}t})^{-1} = \mathrm{e}^{-\boldsymbol{A}t} = \boldsymbol{\Phi}(-t)$

(5) $\boldsymbol{\Phi}(t_2 - t_0) = \boldsymbol{\Phi}(t_2 - t_1)\boldsymbol{\Phi}(t_1 - t_0)$ (2.14)

Proof $\boldsymbol{\Phi}(t_2 - t_1)\boldsymbol{\Phi}(t_1 - t_0) = e^{A(t_2-t_1)}e^{A(t_1-t_0)} = e^{A(t_2-t_0)} = \boldsymbol{\Phi}(t_2 - t_0)$

(6) $[\boldsymbol{\Phi}(t)]^k = \boldsymbol{\Phi}(kt)$, k is positive integer (2.15)

Proof $[\boldsymbol{\Phi}(t)]^k = (e^{At})^k = e^{kAt} = \boldsymbol{\Phi}(kt)$

(7) $e^{(A+B)t} = e^{At} \cdot e^{Bt}$ if and only if $AB = BA$ (2.16)

Proof Since

$$e^{(A+B)t} = I_n + (A + B)t + \frac{1}{2!}(A + B)^2t^2 + \frac{1}{3!}(A + B)^3t^3 + \cdots =$$
$$I_n + (A + B)t + \frac{1}{2!}(A^2 + AB + BA + B^2)t^2 +$$
$$\frac{1}{3!}(A^3 + A^2B + ABA + AB^2 + BA^2 + BAB + B^2A + B^3)t^3 + \cdots$$

and

$$e^{At} \cdot e^{Bt} = \left(I_n + At + \frac{1}{2!}A^2t^2 + \frac{1}{3!}A^3t^3 + \cdots\right)\left(I_n + Bt + \frac{1}{2!}B^2t^2 + \frac{1}{3!}B^3t^3 + \cdots\right) =$$
$$I_n + (A + B)t + \frac{1}{2!}(A^2 + 2AB + B^2)t^2 + \frac{1}{3!}(A^3 + 3A^2B + 3AB^2 + B^3)t^3 + \cdots$$

comparing these two equations yields

$$e^{(A+B)t} - e^{At}e^{Bt} = \frac{1}{2!}(BA - AB)^2t^2 + \frac{1}{3!}(-2A^2B + ABA - 2AB^2 +$$
$$BA^2 + BAB + B^2A)t^3 + \cdots$$

It can be seen, from the resulting equation, that

$$e^{(A+B)t} = e^{At} \cdot e^{Bt}$$

if and only if

$$AB = BA$$

4. Homogeneous Solution for $x(t_0) = x_0$

Now, we derive the homogeneous solution if the initial state is defined at some general time $t_0 \neq 0$, i.e., $x(t_0) = x_0$. Since

$$x(t_0) = \boldsymbol{\Phi}(t_0)x(0)$$

hence

$$x(t) = \boldsymbol{\Phi}(t)x(0) = \boldsymbol{\Phi}(t)\boldsymbol{\Phi}^{-1}(t_0)x(t_0) = \boldsymbol{\Phi}(t)\boldsymbol{\Phi}(-t_0)x(t_0) =$$
$$\boldsymbol{\Phi}(t - t_0)x(t_0) = \boldsymbol{\Phi}(t - t_0)x_0 =$$
$$e^{A(t-t_0)}x(t_0) = e^{A(t-t_0)}x_0 \tag{2.17}$$

2.1.2 Calculation of e^{At}

1. Using Infinite Series of e^{At}

The matrix exponential function e^{At} can be calculated by using the infinite series in Eq. (2.7)

$$e^{At} \approx \sum_{k=0}^{N} \frac{1}{k!}A^kt^k \tag{2.18}$$

where N is a finite integer. This series will not, generally, give a closed-form solution and is mainly used in computer computation.

2. Using Laplace Transform

Taking Laplace transform of Eq. (2.1) yields

$$sX(s) - x(0) = AX(s)$$

or

$$X(s) = (sI - A)^{-1}x(0)$$

Comparing this equation with the solution in Eq. (2.8), we have

$$e^{At} = \mathscr{L}^{-1}[(sI - A)^{-1}] \tag{2.19}$$

Example 2.1 Calculate the matrix exponential function e^{At} if

$$A = \begin{bmatrix} 0 & 1 \\ -2 & -3 \end{bmatrix}$$

Solution To find the matrix exponential function we first calculate the matrix $(sI - A)^{-1}$

$$(sI - A)^{-1} = \begin{bmatrix} s & -1 \\ 2 & s+3 \end{bmatrix}^{-1} = \frac{\operatorname{adj}(sI - A)}{\det(sI - A)} = \frac{1}{(s+1)(s+2)}\begin{bmatrix} s+3 & 1 \\ -2 & s \end{bmatrix} =$$

$$\begin{bmatrix} \dfrac{2}{s+1} - \dfrac{1}{s+2} & \dfrac{1}{s+1} - \dfrac{1}{s+2} \\ -\dfrac{2}{s+1} + \dfrac{2}{s+2} & -\dfrac{1}{s+1} + \dfrac{2}{s+2} \end{bmatrix}$$

The matrix exponential is the inverse Laplace transform of $(sI - A)^{-1}$, i.e.,

$$e^{At} = \mathscr{L}^{-1}[(sI - A)^{-1}] = \begin{bmatrix} 2e^{-t} - e^{-2t} & e^{-t} - e^{-2t} \\ -2e^{-t} + 2e^{-2t} & -e^{-t} + 2e^{-2t} \end{bmatrix}$$

3. Using Jordan Canonical Form

Since the exponential function of a matrix in Jordan form is simple, the similarity transformation, through which an arbitrary given system matrix can be transformed into Jordan form, is an important approach to the calculation of e^{At}.

(1) Matrix exponential of Jordan matrix

Case 1 Matrix exponential of a diagonal matrix

If a $n \times n$ matrix is diagonal, i.e.,

$$\Lambda = \begin{bmatrix} \lambda_1 & & & \\ & \lambda_2 & & \\ & & \ddots & \\ & & & \lambda_n \end{bmatrix} \tag{2.20}$$

then its matrix exponential still is a $n \times n$ diagonal matrix, and is given by

$$e^{At} = \begin{bmatrix} e^{\lambda_1 t} & & & \\ & e^{\lambda_2 t} & & \\ & & \ddots & \\ & & & e^{\lambda_n t} \end{bmatrix} \tag{2.21}$$

Case 2 Matrix exponential of a general Jordan matrix

Assume that a $n \times n$ Jordan matrix $\boldsymbol{J}$ has m distinct eigenvalues $\lambda_1, \cdots, \lambda_i, \cdots, \lambda_m$, and the multiplicity of λ_i is σ_i, i.e.,

$$\boldsymbol{J} = \begin{bmatrix} \boldsymbol{J}_1 & \cdots & 0 & \cdots & 0 \\ \vdots & & \vdots & & \vdots \\ 0 & \cdots & \boldsymbol{J}_i & \cdots & 0 \\ \vdots & & \vdots & & \vdots \\ 0 & \cdots & 0 & \cdots & \boldsymbol{J}_m \end{bmatrix}_{n \times n} \tag{2.22}$$

and the Jordan block

$$\boldsymbol{J}_i = \begin{bmatrix} \lambda_i & 1 & \cdots & 0 & 0 \\ 0 & \lambda_i & \cdots & 0 & 0 \\ \vdots & \vdots & & \vdots & \vdots \\ 0 & 0 & \cdots & \lambda_i & 1 \\ 0 & 0 & \cdots & 0 & \lambda_i \end{bmatrix}_{\sigma_i \times \sigma_i} \tag{2.23}$$

is a $\sigma_i \times \sigma_i$ block. Then, the corresponding matrix exponential is given by

$$e^{\boldsymbol{J}t} = \begin{bmatrix} e^{\boldsymbol{J}_1 t} & \cdots & 0 & \cdots & 0 \\ \vdots & & \vdots & & \vdots \\ 0 & \cdots & e^{\boldsymbol{J}_i t} & \cdots & 0 \\ \vdots & & \vdots & & \vdots \\ 0 & \cdots & 0 & \cdots & e^{\boldsymbol{J}_m t} \end{bmatrix} \tag{2.24}$$

where

$$e^{\boldsymbol{J}_i t} = \boldsymbol{I} + \boldsymbol{J}_i t + \frac{1}{2!}\boldsymbol{J}_i^2 t^2 + \cdots = \begin{bmatrix} 1 & t & \frac{1}{2!}t^2 & \cdots & \frac{1}{(\sigma_i - 1)!}t^{\sigma_i - 1} \\ 0 & 1 & t & \cdots & \frac{1}{(\sigma_i - 2)!}t^{\sigma_i - 2} \\ 0 & 0 & 1 & \cdots & \frac{1}{(\sigma_i - 3)!}t^{\sigma_i - 3} \\ \vdots & \vdots & & & \vdots \\ 0 & 0 & 0 & \cdots & 1 \end{bmatrix} e^{\lambda_i t} \tag{2.25}$$

Eq. (2.20) and Eq. (2.25) can easily be proven by using the power series expansion of the matrix exponential function.

(2) Calculation of e^{At} by using Jordan form

From the matrix theory, for a given square matrix $\boldsymbol{A}$, there exists a nonsingular constant matrix $\boldsymbol{P}$, such that $\bar{\boldsymbol{A}} = \boldsymbol{P}^{-1}\boldsymbol{A}\boldsymbol{P}$. Then, we have

$$\mathrm{e}^{\bar{A}t} = \mathrm{e}^{(P^{-1}AP)t} = I + P^{-1}APt + \frac{1}{2!}(P^{-1}AP)^2t^2 + \cdots + \frac{1}{k!}(P^{-1}AP)^k t^k + \cdots =$$

$$P^{-1}IP + P^{-1}APt + \frac{1}{2!}P^{-1}A^2Pt^2 + \cdots + \frac{1}{k!}P^{-1}A^kPt^k + \cdots =$$

$$P^{-1}[I + P^{-1}APt + \frac{1}{2!}(P^{-1}AP)^2t^2 + \cdots]P = P^{-1}\mathrm{e}^{At}P$$

Thus, the matrix exponential function of a system matrix $\boldsymbol{A}$ can be given by

$$\mathrm{e}^{At} = P\mathrm{e}^{(P^{-1}AP)t}P^{-1} \tag{2.26}$$

Especially, if $\boldsymbol{P}^{-1}\boldsymbol{A}\boldsymbol{P}$ is of Jordan form, including the diagonal form, then the matrix exponential e^{At} can be easily obtained from Eq. (2.21), Eq. (2.24) and Eq. (2.25).

(3) Transforming a square matrix into Jordan form

Case 1 All the eigenvalues of $\boldsymbol{A}$ are distinct

We assume that all the eigenvalues, $\lambda_1, \lambda_2, \cdots, \lambda_n$, of a $n \times n$ matrix $\boldsymbol{A}$ are distinct. In this case there exists a similarity transformation $\boldsymbol{P}^{-1}\boldsymbol{A}\boldsymbol{P} = \boldsymbol{\Lambda}$ so that the matrix $\boldsymbol{\Lambda}$ is of diagonal form.

The $\boldsymbol{P}$ matrix can be formed by use of the linear independent eigenvectors of $\boldsymbol{A}$, i.e.,

$$P = [p_1 \quad p_2 \quad \cdots \quad p_n] \tag{2.27}$$

where $\boldsymbol{p}_i (i = 1,2,\cdots,n)$ denotes the eigenvector that is associated with the eigenvalue λ_i. Any nonzero vector $\boldsymbol{p}$ satisfying $\boldsymbol{A}\boldsymbol{p} = \lambda\boldsymbol{p}$ is called an eigenvector of $\boldsymbol{A}$ associated with the eigenvalue λ. The eigenvector $\boldsymbol{p}_i$ associated with the eigenvalue λ_i is given by

$$(\lambda_i I - A)p_i = \mathbf{0}, \ i = 1,2,\cdots,n \tag{2.28}$$

Particularly, if $\boldsymbol{A}$ is a companion matrix, it can be shown that $\boldsymbol{P}$ may be the Vandermonde matrix, i.e.,

$$P = \begin{bmatrix} 1 & 1 & \cdots & 1 \\ \lambda_1 & \lambda_2 & \cdots & \lambda_n \\ \vdots & \vdots & & \vdots \\ \lambda_1^{n-1} & \lambda_2^{n-1} & \cdots & \lambda_n^{n-1} \end{bmatrix} \tag{2.29}$$

where $\lambda_1, \lambda_2, \cdots, \lambda_n$ are the eigenvalues of $\boldsymbol{A}$.

Example 2.2 Again consider the matrix given in Example 2.1, i.e.,

$$A = \begin{bmatrix} 0 & 1 \\ -2 & -3 \end{bmatrix}$$

Calculate the matrix exponential function, using the similarity transformation.

Solution Since

$$\det(sI - A) = \begin{vmatrix} s & -1 \\ 2 & s+3 \end{vmatrix} = s^2 + 3s + 2 = (s+1)(s+2)$$

The eigenvalues are $\lambda_1 = -1$ and $\lambda_2 = -2$. By inspection, the matrix $\boldsymbol{A}$ is a companion matrix and the similarity transformation may be carried on by use of the Vandermonde matrix, i.e.,

$$P = \begin{bmatrix} 1 & 1 \\ \lambda_1 & \lambda_2 \end{bmatrix} = \begin{bmatrix} 1 & 1 \\ -1 & -2 \end{bmatrix}$$

The inverse of P is given by

$$P^{-1} = \frac{\text{adj}P}{\det P} = \begin{bmatrix} 2 & 1 \\ -1 & -1 \end{bmatrix}$$

Hence, the diagonal form is now obtained as

$$\Lambda = P^{-1}AP = \begin{bmatrix} -1 & 0 \\ 0 & -2 \end{bmatrix}$$

Finally, from Eq. (2.26), the matrix exponential function is

$$e^{At} = Pe^{(P^{-1}AP)t}P^{-1} = Pe^{\Lambda t}P^{-1} = \begin{bmatrix} 1 & 1 \\ -1 & -2 \end{bmatrix}\begin{bmatrix} e^{-t} & 0 \\ 0 & e^{-2t} \end{bmatrix}\begin{bmatrix} 2 & 1 \\ -1 & -1 \end{bmatrix} =$$

$$\begin{bmatrix} 2e^{-t} - e^{-2t} & e^{-t} - e^{-2t} \\ -2e^{-t} + 2e^{-2t} & -e^{-t} + 2e^{-2t} \end{bmatrix}$$

Case 2 The eigenvalues of A are not all distinct.

When a matrix A has multiple eigenvalues, the matrix can be diagonalized if the number of independent eigenvectors associated to each multiple-eigenvalue is equal to the multiplicity of the eigenvalue.

Example 2.3 Diagonalizing the following matrix

$$A = \begin{bmatrix} 1 & 0 & -1 \\ 0 & 1 & 0 \\ 0 & 0 & 2 \end{bmatrix}$$

Solution The eigenvalues of A are $\lambda_1 = \lambda_2 = 1$, $\lambda_3 = 2$. The eigenvectors associated with $\lambda = 1$ can be found by solving the equation

$$(\lambda_1 I - A)p = \begin{bmatrix} 0 & 0 & -1 \\ 0 & 0 & 0 \\ 0 & 0 & -1 \end{bmatrix} p = \mathbf{0}$$

Note that two linearly independent eigenvectors, say $p_1 = [1 \quad 0 \quad 0]^T$ and $p_2 = [0 \quad 1 \quad 0]^T$, associated with $\lambda_1 = \lambda_2 = 1$ can be found for this equation. Solving the equation

$$(\lambda_3 I - A)p = \begin{bmatrix} 1 & 0 & 1 \\ 0 & 1 & 0 \\ 0 & 0 & 0 \end{bmatrix} p = \mathbf{0}$$

an eigenvector associated with $\lambda_3 = 2$ can be found as $p_3 = [1 \quad 0 \quad -1]^T$. Hence we obtain

$$P = [p_1 \quad p_2 \quad p_3] = \begin{bmatrix} 1 & 0 & 1 \\ 0 & 1 & 0 \\ 0 & 0 & -1 \end{bmatrix}, \quad P^{-1} = \frac{\text{adj}P}{\det P} = \begin{bmatrix} 1 & 0 & 1 \\ 0 & 1 & 0 \\ 0 & 0 & -1 \end{bmatrix}$$

The matrix $P^{-1}AP$ is estimated and found to be

$$\boldsymbol{P}^{-1}\boldsymbol{A}\boldsymbol{P} = \begin{bmatrix} 1 & 0 & 0 \\ 0 & 1 & 0 \\ 0 & 0 & 2 \end{bmatrix} = \boldsymbol{\Lambda}$$

In this example, the matrix having repeated eigenvalues can be diagonalized. Generally, however, it is not always possible to diagonalize the matrix $\boldsymbol{A}$ having multiple eigenvalues. In that case the matrix can be transformed only into a Jordan form by use of a similarity transformation, i.e., $\boldsymbol{P}^{-1}\boldsymbol{A}\boldsymbol{P} = \boldsymbol{J}$.

When a matrix $\boldsymbol{A}$ has multiple eigenvalues, the number of eigenvectors associated with a multiple eigenvalue is often less than the multiplicity of the eigenvalue, hence more vectors are needed to form a transformation matrix. Such vectors are related to the corresponding eigenvector and called the generalized eigenvectors. For example, if the matrix $\boldsymbol{A}$ has $\lambda_1 = \lambda_2 = \lambda_3$ and there is only one eigenvector, denoted by $\boldsymbol{p}_1$, associated with this eigenvalue, then we can obtain two generalized eigenvectors, denoted by $\boldsymbol{p}_2$ and $\boldsymbol{p}_3$, from the equations

$$(\lambda_1\boldsymbol{I} - \boldsymbol{A})\boldsymbol{p}_2 = -\boldsymbol{p}_1$$
$$(\lambda_1\boldsymbol{I} - \boldsymbol{A})\boldsymbol{p}_3 = -\boldsymbol{p}_2$$

Example 2.4 Transform the following matrix into a Jordan form

$$\boldsymbol{A} = \begin{bmatrix} 1 & 1 & 2 \\ 0 & 1 & 3 \\ 0 & 0 & 2 \end{bmatrix}$$

Solution The eigenvalues of $\boldsymbol{A}$ are $\lambda_1 = \lambda_2 = 1$, $\lambda_3 = 2$. The eigenvector associated with $\lambda = 1$ can be found by solving the equation

$$(\lambda_1\boldsymbol{I} - \boldsymbol{A})\boldsymbol{p} = \begin{bmatrix} 0 & -1 & -2 \\ 0 & 0 & -3 \\ 0 & 0 & -1 \end{bmatrix}\boldsymbol{p} = \boldsymbol{0}$$

It can be seen that we can find only one linearly independent eigenvector, say $\boldsymbol{p}_1 = [1 \ \ 0 \ \ 0]^{\mathrm{T}}$, associated with $\lambda_1 = \lambda_2 = 1$. Another vector, associated with $\lambda_1 = \lambda_2 = 1$, needed to form a transformation matrix, can be found by solving the equation

$$(\lambda_1\boldsymbol{I} - \boldsymbol{A})\boldsymbol{p}_2 = -\boldsymbol{p}_1$$

i.e.,

$$\begin{bmatrix} 0 & -1 & -2 \\ 0 & 0 & -3 \\ 0 & 0 & -1 \end{bmatrix}\boldsymbol{p}_2 = -\begin{bmatrix} 1 \\ 0 \\ 0 \end{bmatrix}$$

Clearly, $\boldsymbol{p}_2 = [0 \ \ 1 \ \ 0]^{\mathrm{T}}$ is a vector solution. Solving the equation

$$(\lambda_3\boldsymbol{I} - \boldsymbol{A})\boldsymbol{p}_3 = \boldsymbol{0}$$

yields the eigenvector associated with $\lambda_3 = 2$ as $\boldsymbol{p}_3 = [5 \ \ 3 \ \ 1]^{\mathrm{T}}$. Hence we have

$$\boldsymbol{P} = \begin{bmatrix} 1 & 0 & 5 \\ 0 & 1 & 3 \\ 0 & 0 & 1 \end{bmatrix}, \quad \boldsymbol{P}^{-1} = \begin{bmatrix} 1 & 0 & -5 \\ 0 & 1 & -3 \\ 0 & 0 & 1 \end{bmatrix}$$

The result of transformation is

$$P^{-1}AP = \begin{bmatrix} 1 & 1 & 0 \\ 0 & 1 & 0 \\ 0 & 0 & 2 \end{bmatrix} = J$$

2.1.3 Solution to Inhomogeneous Equation

Now, let's consider the general form of the solution to the state equation

$$\dot{\boldsymbol{x}}(t) = \boldsymbol{A}\boldsymbol{x}(t) + \boldsymbol{B}\boldsymbol{u}(t),\ t \geqslant 0;\ \boldsymbol{x}(0) = \boldsymbol{x}_0 \tag{2.30}$$

The complete solution to the state equation can be derived by two methods, respectively.

1. Integration Method

From Eq. (2.30) we get

$$\mathrm{e}^{-\boldsymbol{A}t}(\dot{\boldsymbol{x}} - \boldsymbol{A}\boldsymbol{x}) = \mathrm{e}^{-\boldsymbol{A}t}\boldsymbol{B}\boldsymbol{u}$$

Since

$$\frac{\mathrm{d}}{\mathrm{d}t}[\mathrm{e}^{-\boldsymbol{A}t}\boldsymbol{x}] = \mathrm{e}^{-\boldsymbol{A}t}\dot{\boldsymbol{x}} + \mathrm{e}^{-\boldsymbol{A}t}(-\boldsymbol{A})\boldsymbol{x} = \mathrm{e}^{-\boldsymbol{A}t}[\dot{\boldsymbol{x}} - \boldsymbol{A}\boldsymbol{x}]$$

so

$$\frac{\mathrm{d}}{\mathrm{d}t}[\mathrm{e}^{-\boldsymbol{A}t}\boldsymbol{x}] = \mathrm{e}^{-\boldsymbol{A}t}\boldsymbol{B}\boldsymbol{u}$$

Integrating this equation yields

$$\mathrm{e}^{-\boldsymbol{A}t}\boldsymbol{x}(t) - \boldsymbol{x}(0) = \int_0^t \mathrm{e}^{-\boldsymbol{A}\tau}\boldsymbol{B}\boldsymbol{u}(\tau)\mathrm{d}\tau$$

and the complete solution is given by

$$\begin{aligned}\boldsymbol{x}(t) &= \mathrm{e}^{\boldsymbol{A}t}\boldsymbol{x}(0) + \int_0^t \mathrm{e}^{\boldsymbol{A}(t-\tau)}\boldsymbol{B}\boldsymbol{u}(\tau)\mathrm{d}\tau = \\ &\boldsymbol{\Phi}(t)\boldsymbol{x}(0) + \int_0^t \boldsymbol{\Phi}(t-\tau)\boldsymbol{B}\boldsymbol{u}(\tau)\mathrm{d}\tau\end{aligned} \tag{2.31}$$

If $t_0 \neq 0$ then the complete solution is given by

$$\begin{aligned}\boldsymbol{x}(t) &= \mathrm{e}^{\boldsymbol{A}(t-t_0)}\boldsymbol{x}(t_0) + \int_{t_0}^t \mathrm{e}^{\boldsymbol{A}(t-\tau)}\boldsymbol{B}\boldsymbol{u}(\tau)\mathrm{d}\tau = \\ &\boldsymbol{\Phi}(t-t_0)\boldsymbol{x}(t_0) + \int_{t_0}^t \boldsymbol{\Phi}(t-\tau)\boldsymbol{B}\boldsymbol{u}(\tau)\mathrm{d}\tau\end{aligned} \tag{2.32}$$

2. Laplace Transform Method

Again consider state equation

$$\dot{\boldsymbol{x}}(t) = \boldsymbol{A}\boldsymbol{x}(t) + \boldsymbol{B}\boldsymbol{u}(t),\ t \geqslant 0$$

Taking Laplace transform yields

$$s\boldsymbol{X}(s) - \boldsymbol{x}(0) = \boldsymbol{A}\boldsymbol{X}(s) + \boldsymbol{B}\boldsymbol{u}(s)$$

or

$$(s\boldsymbol{I} - \boldsymbol{A})\boldsymbol{X}(s) = \boldsymbol{x}(0) + \boldsymbol{B}\boldsymbol{u}(s)$$

Hence

$$x(t) = \mathscr{L}^{-1}[(sI - A)^{-1}]x(0) + \mathscr{L}^{-1}[(sI - A)^{-1}Bu(s)] \tag{2.33}$$

Example 2.5 Find the solution to the state equation given by

$$\dot{x}(t) = \begin{bmatrix} 0 & 1 \\ -2 & -3 \end{bmatrix} x(t) + \begin{bmatrix} 0 \\ 1 \end{bmatrix} u(t),\ u(t) = 1(t),\ x(0) = \begin{bmatrix} x_1(0) \\ x_2(0) \end{bmatrix}$$

Solution (1) Using integration method. Since $u(t) = 1(t)$, letting $\xi = t - \tau$ yields

$$x(t) = \Phi(t)x(0) + \int_0^t \Phi(t-\tau)Bu(\tau)\mathrm{d}\tau = \Phi(t)x(0) - \int_t^0 \Phi(\xi)B\mathrm{d}\xi = \Phi(t)x(0) + \int_0^t \Phi(\tau)B\mathrm{d}\tau$$

From Example 2.1 we get

$$\Phi(t) = \begin{bmatrix} 2e^{-t} - e^{-2t} & e^{-t} - e^{-2t} \\ -2e^{-t} + 2e^{-2t} & -e^{-t} + 2e^{-2t} \end{bmatrix}$$

Hence

$$\int_0^t \Phi(\tau)B\mathrm{d}\tau = \int_0^t \begin{bmatrix} 2e^{-\tau} - e^{-2\tau} & e^{-\tau} - e^{-2\tau} \\ -2e^{-\tau} + 2e^{-2\tau} & -e^{-\tau} + 2e^{-2\tau} \end{bmatrix}\begin{bmatrix} 0 \\ 1 \end{bmatrix}\mathrm{d}\tau = \int_0^t \begin{bmatrix} e^{-\tau} - e^{-2\tau} \\ -e^{-\tau} + 2e^{-2\tau} \end{bmatrix}\mathrm{d}\tau =$$

$$\left.\begin{bmatrix} -e^{-\tau} + \frac{1}{2}e^{-2\tau} \\ e^{-\tau} - e^{-2\tau} \end{bmatrix}\right|_0^t = \begin{bmatrix} \frac{1}{2} - e^{-t} + \frac{1}{2}e^{-2t} \\ e^{-t} - e^{-2t} \end{bmatrix}$$

$$x(t) = \begin{bmatrix} 2e^{-t} - e^{-2t} & e^{-t} - e^{-2t} \\ -2e^{-t} + 2e^{-2t} & -e^{-t} + 2e^{-2t} \end{bmatrix}\begin{bmatrix} x_1(0) \\ x_2(0) \end{bmatrix} + \begin{bmatrix} \frac{1}{2} - e^{-t} + \frac{1}{2}e^{-2t} \\ e^{-t} - e^{-2t} \end{bmatrix}$$

(2) Using Laplace-transform method. Again, from Example 2.1 we get

$$(sI - A)^{-1} = \begin{bmatrix} \frac{2}{s+1} - \frac{1}{s+2} & \frac{1}{s+1} - \frac{1}{s+2} \\ -\frac{2}{s+1} + \frac{2}{s+2} & -\frac{1}{s+1} + \frac{2}{s+2} \end{bmatrix}$$

$$\mathscr{L}^{-1}[(sI - A)^{-1}] = \begin{bmatrix} 2e^{-t} - e^{-2t} & e^{-t} - e^{-2t} \\ -2e^{-t} + 2e^{-2t} & -e^{-t} + 2e^{-2t} \end{bmatrix}$$

Thus

$$(sI - A)^{-1}BU(s) = \begin{bmatrix} \frac{2}{s+1} - \frac{1}{s+2} & \frac{1}{s+1} - \frac{1}{s+2} \\ -\frac{2}{s+1} + \frac{2}{s+2} & -\frac{1}{s+1} + \frac{2}{s+2} \end{bmatrix}\begin{bmatrix} 0 \\ 1 \end{bmatrix}\cdot\frac{1}{s} =$$

$$\begin{bmatrix} \frac{1}{s+1} - \frac{1}{s+2} \\ -\frac{1}{s+1} + \frac{2}{s+2} \end{bmatrix}\cdot\frac{1}{s} = \begin{bmatrix} \frac{1}{s(s+1)(s+2)} \\ \frac{1}{(s+1)(s+2)} \end{bmatrix} =$$

$$\begin{bmatrix} \frac{1}{2}\cdot\frac{1}{s} - \frac{1}{s+1} + \frac{1}{2}\cdot\frac{1}{s+2} \\ \frac{1}{s+1} - \frac{1}{s+2} \end{bmatrix}$$

From Eq. (2.33) we have

$$\boldsymbol{x}(t) = \mathscr{L}^{-1}[(s\boldsymbol{I} - \boldsymbol{A})^{-1}]\boldsymbol{x}(0) + \mathscr{L}^{-1}[(s\boldsymbol{I} - \boldsymbol{A})^{-1}\boldsymbol{Bu}(s)] =$$

$$\begin{bmatrix} 2e^{-t} - e^{-2t} & e^{-t} - e^{-2t} \\ -2e^{-t} + 2e^{-2t} & -e^{-t} + 2e^{-2t} \end{bmatrix} \begin{bmatrix} x_1(0) \\ x_2(0) \end{bmatrix} + \begin{bmatrix} \frac{1}{2} - e^{-t} + \frac{1}{2}e^{-2t} \\ e^{-t} - e^{-2t} \end{bmatrix}$$

2.2 Time-Invariant Discrete Systems

2.2.1 State-Space Representation

Linear discrete-time systems are usually described by the difference equation or the pulse transfer function. The general form of difference equation for a single variable system is

$$y(k+n) + a_1 y(k+n-1) + \cdots + a_{n-1}y(k+1) + a_n y(k) = b_1 u(k+n-1) + \cdots + b_{n-1}u(k+1) + b_n u(k) \tag{2.34}$$

where $u(k)$ and $y(k)$ are the input and output, respectively, at the kth sampling instant. In a similar way as that for the continuous-time case, we can establish the corresponding canonical state-space representation.

For example, if the state variables are chosen as

$$x_1(k) = z(k)$$
$$x_2(k) = z(k+1)$$
$$\vdots$$
$$x_n(k) = z(k+n-1)$$

for which the equation $z(k+n) + a_1 z(k+n-1) + \cdots + a_n z(k) = u(k)$ is satisfied, then the state-space representation is of the controllable canonical form

$$\left.\begin{aligned} \boldsymbol{x}(k+1) &= \begin{bmatrix} 0 & & \\ \vdots & & \boldsymbol{I}_{n-1} \\ 0 & & \\ -a_n & -a_{n-1} & \cdots & -a_1 \end{bmatrix} \boldsymbol{x}(k) + \begin{bmatrix} 0 \\ \vdots \\ 0 \\ 1 \end{bmatrix} u(k) \\ y(k) &= [b_n \quad b_{n-1} \quad \cdots \quad b_1]\boldsymbol{x}(k) \end{aligned}\right\} \tag{2.35}$$

The standard form of the state-space representation for the discrete systems is denoted as

$$\left.\begin{aligned} \boldsymbol{x}(k+1) &= \boldsymbol{Ax}(k) + \boldsymbol{Bu}(k) \\ \boldsymbol{y}(k) &= \boldsymbol{Cx}(k) + \boldsymbol{Du}(k) \end{aligned}\right\} \tag{2.36}$$

where the matrix $\boldsymbol{D}$ usually is zero. Note that a discrete-time state equation is a set of first-order difference equations instead of a set of differential equation as in the continuous-time case.

2.2.2 Discretization of Continuous-time State Equation

A continuous-time system can also be modeled as a discrete-time system if its response is of interested, or measured, only at certain instants of time. For convenience, the discrete instants of time, at which the output appear, will be assumed equally spaced and the interval is denoted by T.

1. Discretization

The state equation of a linear time-invariant continuous system is usually written as

$$\dot{\boldsymbol{x}}(t) = \boldsymbol{A}\boldsymbol{x}(t) + \boldsymbol{B}\boldsymbol{u}(t)$$

and its solution is

$$\boldsymbol{x}(t) = \mathrm{e}^{\boldsymbol{A}(t-t_0)}\boldsymbol{x}(t_0) + \int_{t_0}^{t} \mathrm{e}^{\boldsymbol{A}(t-\tau)}\boldsymbol{B}\boldsymbol{u}(\tau)\mathrm{d}\tau$$

Taking $t_0 = kT$ and $t = (k+1)T$, we have

$$\boldsymbol{x}[(k+1)T] = \mathrm{e}^{\boldsymbol{A}T}\boldsymbol{x}(kT) + \int_{kT}^{(k+1)T} \mathrm{e}^{\boldsymbol{A}[(k+1)T-\tau]}\boldsymbol{B}\boldsymbol{u}(\tau)\mathrm{d}\tau$$

Letting $\tau = kT + \xi$, where $0 \leqslant \xi \leqslant T$, and substituting it into above equation, yield

$$\boldsymbol{x}[(k+1)T] = \mathrm{e}^{\boldsymbol{A}T}x(kT) + \int_0^T \mathrm{e}^{\boldsymbol{A}(T-\xi)}\boldsymbol{B}\boldsymbol{u}(kT+\xi)\mathrm{d}\xi \approx$$

$$\mathrm{e}^{\boldsymbol{A}T}\boldsymbol{x}(kT) + \int_0^T \mathrm{e}^{\boldsymbol{A}(T-\xi)}\boldsymbol{B}\mathrm{d}\xi \cdot \boldsymbol{u}(kT)$$

or

$$\boldsymbol{x}[(k+1)T] = \boldsymbol{A}^*\boldsymbol{x}(kT) + \boldsymbol{B}^*\boldsymbol{u}(kT) \tag{2.37}$$

where

$$\boldsymbol{A}^* = \mathrm{e}^{\boldsymbol{A}T} = \mathrm{e}^{\boldsymbol{A}t}\big|_{t=T} \tag{2.38}$$

$$\boldsymbol{B}^* = \int_0^T \mathrm{e}^{\boldsymbol{A}(T-\xi)}\boldsymbol{B}\mathrm{d}\xi = -\int_T^0 \mathrm{e}^{\boldsymbol{A}t}\boldsymbol{B}\mathrm{d}t = \int_0^T \mathrm{e}^{\boldsymbol{A}t}\boldsymbol{B}\mathrm{d}t \tag{2.39}$$

As an algebraic equation, the discretized output equation is

$$\boldsymbol{y}(k) = \boldsymbol{C}\boldsymbol{x}(k) + \boldsymbol{D}\boldsymbol{u}(k) \tag{2.40}$$

Example 2.6 Determine the discretization of the following continuous-time state equation

$$\dot{\boldsymbol{x}}(t) = \begin{bmatrix} 0 & 1 \\ -2 & -3 \end{bmatrix}\boldsymbol{x}(t) + \begin{bmatrix} 0 \\ 1 \end{bmatrix}\boldsymbol{u}(t)$$

It is assumed that $T = 1$ s.

Solution from Example 2.1 we have

$$\mathrm{e}^{\boldsymbol{A}t} = \begin{bmatrix} 2\mathrm{e}^{-t} - \mathrm{e}^{-2t} & \mathrm{e}^{-t} - \mathrm{e}^{-2t} \\ -2\mathrm{e}^{-t} + 2\mathrm{e}^{-2t} & -\mathrm{e}^{-t} + 2\mathrm{e}^{-2t} \end{bmatrix}$$

Using Eq. (2.38) and (2.39) yields

$$\boldsymbol{A}^* = \mathrm{e}^{\boldsymbol{A}t}\big|_{t=T} = \begin{bmatrix} 2\mathrm{e}^{-t} - \mathrm{e}^{-2t} & \mathrm{e}^{-t} - \mathrm{e}^{-2t} \\ -2\mathrm{e}^{-t} + 2\mathrm{e}^{-2t} & -\mathrm{e}^{-t} + 2\mathrm{e}^{-2t} \end{bmatrix}\Bigg|_{t=1} = \begin{bmatrix} 0.600\,4 & 0.232\,5 \\ -0.465\,1 & -0.097\,2 \end{bmatrix}$$

$$\boldsymbol{B}^* = \int_0^T \mathrm{e}^{\boldsymbol{A}t}\boldsymbol{B}\mathrm{d}t = \int_0^1 \begin{bmatrix} 2\mathrm{e}^{-t} - \mathrm{e}^{-2t} & \mathrm{e}^{-t} - \mathrm{e}^{-2t} \\ -2\mathrm{e}^{-t} + 2\mathrm{e}^{-2t} & -\mathrm{e}^{-t} + 2\mathrm{e}^{-2t} \end{bmatrix}\begin{bmatrix} 0 \\ 1 \end{bmatrix}\mathrm{d}t = \int_0^1 \begin{bmatrix} \mathrm{e}^{-t} - \mathrm{e}^{-2t} \\ -\mathrm{e}^{-t} + 2\mathrm{e}^{-2t} \end{bmatrix}\mathrm{d}t =$$

$$\begin{bmatrix} \frac{1}{2} - e^{-1} + \frac{1}{2}e^{-2} \\ e^{-1} - 2e^{-2} \end{bmatrix} = \begin{bmatrix} 0.199\,8 \\ 0.232\,5 \end{bmatrix}$$

The resulting discretized state equation is

$$x(k+1) = \begin{bmatrix} 0.600\,4 & 0.232\,5 \\ -0.465\,1 & -0.097\,2 \end{bmatrix} x(k) + \begin{bmatrix} 0.199\,8 \\ 0.232\,5 \end{bmatrix} u(k)$$

2. Approximation

If the instant of time T is sufficient small compared with the time constants of the system, then

$$A^* = e^{AT} = I + AT + \cdots \approx I + AT$$

and

$$B^* = \int_0^T e^{At} B \mathrm{d}t = \int_0^T e^{At} \mathrm{d}t \cdot B \approx \int_0^T (I + At)\mathrm{d}t \cdot B \approx TB$$

Thus, Eq. (2.37) can be rewritten, approximately, as

$$x[(k+1)T] = (I + TA)x(kT) + TBu(kT) \tag{2.41}$$

2.2.3 Solution to Discrete-Time State Equation

The solution to a discrete-time state equation can be obtained by classical method, z-transform method, or recursion method. Here, in what follows the recursion method will be illustrated with an example.

The basic discrete-time state equation is

$$x(k+1) = Ax(k) + Bu(k)$$

Letting $k = 0, 1, \cdots, k-1$ yields

$$\begin{aligned} x(1) &= Ax(0) + Bu(0) \\ x(2) &= Ax(1) + Bu(1) = A^2x(0) + ABu(0) + Bu(1) \\ x(3) &= Ax(2) + Bu(2) = A^3x(0) + A^2Bu(0) + ABu(1) + Bu(2) \\ &\vdots \\ x(k) &= Ax(k-1) + Bu(k-1) = A^kx(0) + A^{k-1}Bu(0) + \cdots + ABu(k-2) + Bu(k-1) \end{aligned}$$

Then the solution is given by the recursive equation

$$x(k) = A^kx(0) + \sum_{i=0}^{k-1} A^{k-i-1}Bu(i) \tag{2.42}$$

In theoretical studies, Eq. (2.42) is often written as

$$x(k) = \Phi(k)x(0) + \sum_{i=0}^{k-1} \Phi(k-i-1)Bu(i) = \Phi(k)x(0) + \sum_{i=0}^{k-1} \Phi(i)Bu(k-i-1) \tag{2.43}$$

where $\Phi(k) = A^k$ is called the discrete state transition matrix.

Problems

P2.1 Calculate the resolvent matrices, i.e., $(s\boldsymbol{I}-\boldsymbol{A})^{-1}$, and the state transition matrices for the following matrices.

(a) $\boldsymbol{A}=\begin{bmatrix}0 & 1\\0 & 0\end{bmatrix}$

(b) $\boldsymbol{A}=\begin{bmatrix}0 & 1\\-1 & 0\end{bmatrix}$

(c) $\boldsymbol{A}=\begin{bmatrix}1 & 0 & 0\\0 & 1 & 0\\0 & 1 & 2\end{bmatrix}$

P2.2 Transform the representations of the following systems into the Jordan canonical form

(a) $\boldsymbol{A}=\begin{bmatrix}-2 & 1\\-17 & -4\end{bmatrix}$, $\boldsymbol{b}=\begin{bmatrix}4\\8\end{bmatrix}$, $\boldsymbol{c}=[0\ \ 1]$

(b) $\boldsymbol{A}=\begin{bmatrix}0 & 1 & 0\\0 & 0 & 1\\2 & 3 & 0\end{bmatrix}$, $\boldsymbol{b}=\begin{bmatrix}0\\0\\9\end{bmatrix}$, $\boldsymbol{c}=[1\ \ 0\ \ 0]$

(c) $\boldsymbol{A}=\begin{bmatrix}-2 & 2 & -1\\0 & -2 & 0\\1 & -4 & 0\end{bmatrix}$, $\boldsymbol{b}=\begin{bmatrix}0\\0\\16\end{bmatrix}$, $\boldsymbol{c}=[1\ \ 0\ \ 0]$

(d) $\boldsymbol{A}=\begin{bmatrix}0 & 1 & 0\\3 & 0 & 2\\-12 & -7 & -6\end{bmatrix}$, $\boldsymbol{B}=\begin{bmatrix}2 & 3\\1 & 5\\7 & 1\end{bmatrix}$, $\boldsymbol{C}=\begin{bmatrix}1 & 0 & 0\\0 & 1 & 0\end{bmatrix}$

P2.3 (a) Determine the characteristic equation, eigenvalues, and eigenvectors of the matrix

$$\boldsymbol{A}=\begin{bmatrix}0 & 1 & 0 & 0\\0 & 0 & 1 & 0\\0 & 0 & 0 & 1\\1 & 0 & 0 & 0\end{bmatrix}$$

(b) Transform the matrix given in (a) into the Jordan form, finding the transformation matrix.

P2.4 A linear system has the following state transition matrix

$$\boldsymbol{\Phi}(t)=\begin{bmatrix}2e^{-t}-e^{-2t} & -2e^{-t}+2e^{-2t}\\-e^{-t}+e^{-2t} & -e^{-t}+2e^{-2t}\end{bmatrix}$$

Determine the system matrix $\boldsymbol{A}$ of the system.

P2.5 Check whether the following matrices satisfy the conditions for the state transition matrix. If any matrix satisfies the condition, find the corresponding system matrix **A**.

(a) $\boldsymbol{\Phi}(t) = \begin{bmatrix} 2e^{-t} - e^{-2t} & -2e^{-t} + 2e^{-2t} \\ e^{-t} - e^{-2t} & -e^{-t} + 2e^{-2t} \end{bmatrix}$

(b) $\boldsymbol{\Phi}(t) = \begin{bmatrix} 1 & \dfrac{1 - 2e^{-2t}}{2} \\ 0 & e^{-2t} \end{bmatrix}$

(c) $\boldsymbol{\Phi}(t) = \begin{bmatrix} 1 & 0 & 0 \\ 0 & \sin t & \cos t \\ 0 & -\cos t & \sin t \end{bmatrix}$

(d) $\boldsymbol{\Phi}(t) = \begin{bmatrix} e^{-t} & 0 & 0 \\ 0 & e^{-5t} & e^{-5t} \\ 0 & e^{-t} & e^{-5t} \end{bmatrix}$

P2.6 The responses of a second-order system, $\dot{\boldsymbol{x}} = \boldsymbol{A}\boldsymbol{x}$, to two different initial states are

$$\boldsymbol{x}(t) = \begin{bmatrix} e^{-3t} \\ -e^{-3t} \end{bmatrix} \text{ when } \boldsymbol{x}(0) = \begin{bmatrix} 1 \\ -4 \end{bmatrix}$$

$$\boldsymbol{x}(t) = \begin{bmatrix} e^{-2t} \\ -e^{-2t} \end{bmatrix} \text{ when } \boldsymbol{x}(0) = \begin{bmatrix} 2 \\ -1 \end{bmatrix}$$

respectively. Determine the system matrix $\boldsymbol{A}$.

P2.7 Consider the state equation

$$\dot{\boldsymbol{x}} = \begin{bmatrix} 1 & 0 \\ 1 & 1 \end{bmatrix} \boldsymbol{x} + \begin{bmatrix} 1 \\ 1 \end{bmatrix} u$$

Determine the state vector $\boldsymbol{x}(t)$ for $t \geqslant 0$ when the input $u(t) = 1(t)$, using two different methods. It is assumed that the initial state is $\boldsymbol{x}(0) = [1 \quad 0]^{\mathrm{T}}$.

P2.8 A discrete system is described by the difference equation

$$y(k+2) + 3y(k+1) + 2y(k) = 2u(k+1) + 3u(k)$$

(a) Determine the state-space representation in controllable canonical form.

(b) Obtain the system response when the input

$$u(k) = \begin{cases} 1 & k = 0,1 \\ 0 & k \neq 0,1 \end{cases}$$

It is assumed that the initial state is zero.

P2.9 Discretize the following continuous system

$$\dot{\boldsymbol{x}} = \begin{bmatrix} 0 & 1 \\ 0 & 2 \end{bmatrix} \boldsymbol{x} + \begin{bmatrix} 0 \\ 1 \end{bmatrix} u$$

$$y = [1 \quad 0] \boldsymbol{x}$$

It is assumed that the sampling period is $T = 1$ s.

Chapter 3 Controllability and Observability of Linear Systems

3.1 Introduction

System analysis generally considers two parts of problems: quantitative and qualitative. In the qualitative study we are interested in the exact response of the system to certain input and initial conditions. In the quantitative study we are interested in the general properties of a system. Controllability and observability are two qualitative properties of linear control systems and play an important role in both theoretical and practical aspects of modern control theory. The concepts of controllability and observability can be roughly illustrated with an example.

Example 3.1 Consider a bridge circuit shown in Fig. 3.1, where the input is v_i and the output $y = v_C$. The state variables of the system are chosen as the voltage across the capacitor and the current through the inductor.

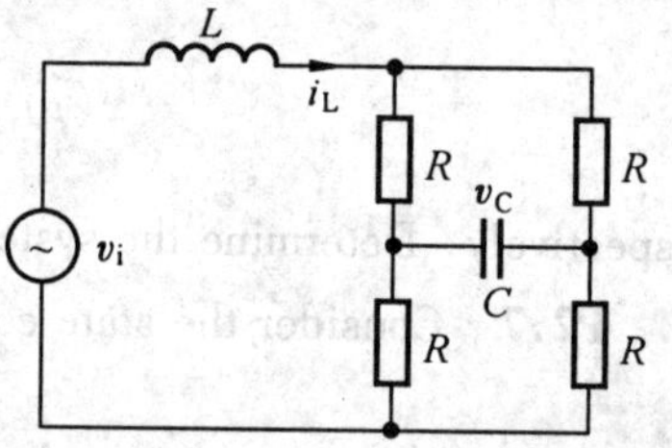

Figure 3.1 Bridge circuit of Example 3.1

Let $x_1 = i_L$ and $x_2 = u_C$. It can be seen from the circuit, if $x_2(t_0) = 0$ then $x_2(t) \equiv 0$ for all $t \geqslant t_0$ no matter what input is applied, i.e., the input has no effect on x_2, and it is said that x_2 is uncontrollable. If the input is zero, no matter what $x_1(t_0)$ is, the change of $x_1(t)$ cannot be reflected in the output y, and it is said that x_1 is unobservable.

3.2 Controllability of Linear Continuous Systems

3.2.1 Definition of Controllability

Roughly speaking, the problem of controllability is to study the possibility of driving the state from the input. If a system is controllable, all the response modes of the system can be excited from the input.

Consider the n-dimensional linear time-invariant continuous system

$$\left.\begin{aligned} \dot{\boldsymbol{x}}(t) &= \boldsymbol{A}\boldsymbol{x}(t) + \boldsymbol{B}\boldsymbol{u}(t) \\ \boldsymbol{y}(t) &= \boldsymbol{C}\boldsymbol{x}(t) + \boldsymbol{D}\boldsymbol{u}(t) \end{aligned}\right\} \tag{3.1}$$

Where $\boldsymbol{A}$, $\boldsymbol{B}$, $\boldsymbol{C}$ and $\boldsymbol{D}$ are $n \times n$, $n \times p$, $q \times n$ and $q \times p$ real constant matrices, respectively. A system is said to be state completely controllable, or to be controllable for short, at time t_0 if there exist a finite time $t_f > t_0$ and an unconstrained control $u(t)$ over $[t_0, t_f]$ which will transfer any nonzero state $\boldsymbol{x}(t_0)$ to $\boldsymbol{x}(t_f) = \boldsymbol{0}$.

This definition requires only that the control $\boldsymbol{u}(t)$ is capable to move the initial state to the object state, what trajectory the state should take is not specified.

3.2.2 Rank Criterion of Controllability

The following rank criterion of state controllability shows that the condition of controllability depends on the coefficient matrices $\boldsymbol{A}$ and $\boldsymbol{B}$ of the system, i.e., the state controllability is a property of the state equation only, the output equation does not play any role here. The criterion also gives one way to examine state controllability.

For the system described by the state equation of Eq. (3.1) to be completely state controllable if and only if the $n \times np$ matrix

$$\boldsymbol{Q}_c = [\boldsymbol{B} \quad \boldsymbol{AB} \quad \cdots \quad \boldsymbol{A}^{n-1}\boldsymbol{B}] \tag{3.2}$$

has rank n. The matrix $\boldsymbol{Q}_c$ is described in terms of $\boldsymbol{A}$ and $\boldsymbol{B}$, and called the controllability matrix.

Proof The solution to the state equation given by Eq. (3.1) is

$$\boldsymbol{x}(t) = e^{\boldsymbol{A}(t-t_0)}\boldsymbol{x}(t_0) + \int_{t_0}^{t} e^{\boldsymbol{A}(t-\tau)}\boldsymbol{Bu}(\tau)d\tau$$

Without losing any generality we can assume that $t_0 = 0$. According to the definition of controllability, if the system is controllable we have

$$\boldsymbol{x}(t_f) = e^{\boldsymbol{A}t_f}\boldsymbol{x}(0) + \int_0^{t_f} e^{\boldsymbol{A}(t_f-\tau)}\boldsymbol{Bu}(\tau)d\tau = 0$$

i.e.,

$$\boldsymbol{x}(0) = -\int_0^{t_f} e^{-\boldsymbol{A}\tau}\boldsymbol{Bu}(\tau)d\tau \tag{3.3}$$

From the Cayley-Hamilton theorem, the matrix exponential function can be written as

$$e^{\boldsymbol{A}t} = \sum_{i=0}^{n-1}\alpha_i(t)\boldsymbol{A}^i \tag{3.4}$$

Substituting Eq. (3.4) into Eq. (3.3) and rearranging yield

$$\boldsymbol{x}(0) = -\int_0^{t_f}\sum_{i=0}^{n-1}\alpha_i(\tau)\boldsymbol{A}^i\boldsymbol{Bu}(\tau)d\tau = -\sum_{i=0}^{n-1}\boldsymbol{A}^i\boldsymbol{B}\int_0^{t_f}\alpha_i(\tau)\boldsymbol{u}(\tau)d\tau \tag{3.5}$$

Let

$$\boldsymbol{f}_i = \int_0^{t_f}\alpha_i(\tau)\boldsymbol{u}(\tau)d\tau \tag{3.6}$$

Then Eq. (3.5) becomes

$$\boldsymbol{x}(0) = -\sum_{i=0}^{n-1}\boldsymbol{A}^i\boldsymbol{B}\boldsymbol{f}_i = -[\boldsymbol{B} \quad \boldsymbol{AB} \quad \cdots \quad \boldsymbol{A}^{n-1}\boldsymbol{B}]\begin{bmatrix}\boldsymbol{f}_0\\ \boldsymbol{f}_1\\ \vdots\\ \boldsymbol{f}_{n-1}\end{bmatrix} = -\boldsymbol{Q}_c\boldsymbol{F}_u \tag{3.7}$$

Now, the controllability may be interpreted as: Given any initial state $\boldsymbol{x}(t_0)$, find the control vector $\boldsymbol{u}(t)$ so that the final state $\boldsymbol{x}(t_f) = \boldsymbol{0}$ for a finite time $t_f > 0$. The problem is to solve $\boldsymbol{F}_u$ from Eq. (3.7) for the given $\boldsymbol{x}(t_0)$ and matrix $\boldsymbol{Q}_c$. Therefore the system is completely state controllable if and only if there exists a set of n linear independent column vector in $\boldsymbol{Q}_c$, or the matrix $\boldsymbol{Q}_c$ has rank n. For a system with one input, the matrix $\boldsymbol{Q}_c$ is square, thus the condition of controllability is that $\boldsymbol{Q}_c$ must be nonsingular.

Since only the matrices $\boldsymbol{A}$ and $\boldsymbol{B}$ are involved for examining the state controllability of a system, sometimes we say that a pair $(\boldsymbol{A}, \boldsymbol{B})$ is controllable, which means that the matrix $\boldsymbol{Q}_c$ has rank n and the system is completely state controllable.

Example 3.2 Examine the state controllability of a system with the state equation given by

$$\dot{\boldsymbol{x}} = \begin{bmatrix} -1 & -2 & -2 \\ 0 & -1 & 1 \\ 1 & 0 & -1 \end{bmatrix} \boldsymbol{x} + \begin{bmatrix} 2 \\ 0 \\ 1 \end{bmatrix} u$$

Solution From the given state equation we get

$$\boldsymbol{AB} = \begin{bmatrix} -1 & -2 & -2 \\ 0 & -1 & 1 \\ 1 & 0 & -1 \end{bmatrix} \begin{bmatrix} 2 \\ 0 \\ 1 \end{bmatrix} = \begin{bmatrix} -4 \\ 1 \\ 1 \end{bmatrix}$$

$$\boldsymbol{A}^2\boldsymbol{B} = \begin{bmatrix} -1 & -2 & -2 \\ 0 & -1 & 1 \\ 1 & 0 & -1 \end{bmatrix} \begin{bmatrix} -4 \\ 1 \\ 1 \end{bmatrix} = \begin{bmatrix} 0 \\ 0 \\ -5 \end{bmatrix}$$

Thus, the controllability matrix is

$$\boldsymbol{Q}_c = [\boldsymbol{B} \quad \boldsymbol{AB} \quad \boldsymbol{A}^2\boldsymbol{B}] = \begin{bmatrix} 2 & -4 & 0 \\ 0 & 1 & 0 \\ 1 & 1 & -5 \end{bmatrix}$$

Since

$$\operatorname{rank} \boldsymbol{Q}_c = 3 = n$$

the system is controllable.

Example 3.3 Determine the condition on parameter a for controllability if the state equation is given by

$$\dot{\boldsymbol{x}} = \begin{bmatrix} -2 & 0 \\ a & -3 \end{bmatrix} \boldsymbol{x} + \begin{bmatrix} 1 \\ 0 \end{bmatrix} u$$

Solution Generating the controllability matrix yields

$$\boldsymbol{Q}_c = \begin{bmatrix} 1 & -2 \\ 0 & a \end{bmatrix}$$

Since the determinant of $\boldsymbol{Q}_c$ is equal to a, which is nonzero only when a is nonzero, the system is controllable for $a \neq 0$.

Example 3.4 Prove that a single-input system with the state equation in controllable canonical form must be controllable.

Proof Denote the state equation be

$$\dot{\boldsymbol{x}} = \begin{bmatrix} 0 & 1 & \cdots & 0 \\ \vdots & \vdots & & \vdots \\ 0 & 0 & \cdots & 1 \\ * & * & * & * \end{bmatrix} \boldsymbol{x} + \begin{bmatrix} 0 \\ \vdots \\ 0 \\ 1 \end{bmatrix} u$$

where * s denote possible nonzero elemensts. The controllability matrix is

$$\boldsymbol{Q}_c = \begin{bmatrix} 0 & 0 & \cdots & 1 \\ \vdots & \vdots & & * \\ 0 & 1 & \cdots & * \\ 1 & * & \cdots & * \end{bmatrix}$$

Obviously, since $\boldsymbol{Q}_c$ is a triangular matrix and it must be nonsingular, the system is controllable.

3.2.3 Controllability Criterion for Jordan Canonical Form

1. All Eigenvalues Are Distinct

If all the eigenvalues of a system matrix are distinct, the state equation of the derived Jordan canonical form is

$$\dot{\boldsymbol{x}} = \boldsymbol{A}\boldsymbol{x} + \boldsymbol{B}u = \begin{bmatrix} \lambda_1 & & & \\ & \lambda_2 & & \\ & & \ddots & \\ & & & \lambda_n \end{bmatrix} \boldsymbol{x} + \boldsymbol{B}\boldsymbol{u}$$

In this case, the matrix pair $(\boldsymbol{A}, \boldsymbol{B})$ is controllable if and only if the matrix $\boldsymbol{B}$ has no rows that are all zeros.

For example, the system described by

$$\dot{\boldsymbol{x}} = \begin{bmatrix} -2 & 0 \\ 0 & -3 \end{bmatrix} \boldsymbol{x} + \begin{bmatrix} 1 \\ 2 \end{bmatrix} u$$

is controllable, and the system described by

$$\dot{\boldsymbol{x}} = \begin{bmatrix} -2 & 0 \\ 0 & -3 \end{bmatrix} \boldsymbol{x} + \begin{bmatrix} 0 \\ 2 \end{bmatrix} u$$

is uncontrollable.

2. Not All Eigenvalues Are Distinct

Assume that the matrices $\boldsymbol{A}$, $\boldsymbol{B}$ and $\boldsymbol{C}$ of derived Jordan form are given as

$$\left.\begin{aligned}
&\underset{(n\times n)}{A} = \begin{bmatrix} J_1 & & & \\ & J_2 & & \\ & & \ddots & \\ & & & J_m \end{bmatrix}, \quad \underset{(n\times p)}{B} = \begin{bmatrix} B_1 \\ B_2 \\ \vdots \\ B_m \end{bmatrix}, \quad \underset{(q\times n)}{C} = [C_1 \quad C_2 \quad \cdots \quad C_m] \\
&\underset{(\sigma_i\times\sigma_i)}{J_i} = \begin{bmatrix} J_{i1} & & & \\ & J_{i2} & & \\ & & \ddots & \\ & & & J_{i\alpha_i} \end{bmatrix}, \quad \underset{(\sigma_i\times p)}{B_i} = \begin{bmatrix} B_{i1} \\ B_{i2} \\ \vdots \\ B_{i\alpha_i} \end{bmatrix}, \quad \underset{(q\times\sigma_i)}{C_i} = [C_{i1} \quad C_{i2} \quad \cdots \quad C_{i\alpha_i}] \\
&\underset{(r_{ij}\times r_{ij})}{J_{ij}} = \begin{bmatrix} \lambda_i & 1 & & \\ & \lambda_i & \ddots & \\ & & \ddots & 1 \\ & & & \lambda_i \end{bmatrix}, \quad \underset{(r_{ij}\times p)}{B_{ij}} = \begin{bmatrix} b_{1ij} \\ b_{2ij} \\ \vdots \\ b_{lij} \end{bmatrix}, \quad \underset{(q\times r_{ij})}{C_{ij}} = [c_{1ij} \quad c_{2ij} \quad \cdots \quad c_{lij}]
\end{aligned}\right\} \tag{3.8}$$

where the $n \times n$ matrix A with m distinct eigenvalues $\lambda_1, \lambda_2, \cdots, \lambda_m$ is in Jordan form, J_i denotes all the Jordan blocks associated with the eigenvalue λ_i, α_i is the number of Jordan blocks in J_i, J_{ij} is the jth Jordan block in J_i, and

$$n = \sum_{i=1}^{m} \sigma_i = \sum_{i=1}^{m}\sum_{j=1}^{\alpha_i} r_{ij}$$

The first row and the last row of B_{ij} are denoted by b_{1ij} and b_{lij}, respectively. The first column and the last column of C_{ij} are denoted by c_{1ij} and c_{lij}, respectively.

In this case, the system is controllable if and only if for each $i = 1, 2, \cdots, m$, the rows of the $\alpha_i \times p$ matrix

$$B_i^l = \begin{bmatrix} b_{li1} \\ b_{li2} \\ \vdots \\ b_{li\alpha_i} \end{bmatrix} \tag{3.9}$$

are linearly independent.

Example 3.5 Consider the controllability of the following systems.

(1) $\dot{x} = \begin{bmatrix} -2 & 0 \\ 0 & -2 \end{bmatrix} x + \begin{bmatrix} 1 \\ 2 \end{bmatrix} u$;

(2) $\dot{x} = \left[\begin{array}{ccc:cc} -2 & & & & \\ & -2 & 1 & & \\ & & -2 & & \\ \hdashline & & & -5 & 1 \\ & & & & -5 \end{array}\right] x + \begin{bmatrix} 1 & 0 \\ 0 & 2 \\ 0 & 1 \\ \hdashline 0 & 0 \\ 2 & 1 \end{bmatrix} u$;

$$(3)\ \dot{\boldsymbol{x}} = \left[\begin{array}{cccc:ccc} -2 & 1 & & & & & \\ & -2 & & & & & \\ & & -2 & & & & \\ & & & -2 & & & \\ \hdashline & & & & 3 & 1 & \\ & & & & & 3 & \\ & & & & & & 3 \end{array}\right]\boldsymbol{x} + \left[\begin{array}{ccc} 0 & 0 & 0 \\ 1 & 0 & 0 \\ 0 & 3 & 0 \\ 0 & 0 & 2 \\ \hdashline 0 & 0 & 0 \\ 1 & 1 & 0 \\ 0 & 2 & 1 \end{array}\right]\boldsymbol{u}.$$

Solution (1)There is only one distinct eigenvalue and $\alpha_1 = -2$. Since

$$\boldsymbol{B}_1^l = \begin{bmatrix} 1 \\ 2 \end{bmatrix}$$

and the two constants in this matrix are always linear dependent, the system is uncontrollable.

(2)Since

$$\boldsymbol{B}_1^l = \begin{bmatrix} 0 & 2 \\ 0 & 1 \end{bmatrix}$$

where the two rows are linear dependent, thus the system is uncontrollable.

(3)From Eq. (3.9) we have

$$\boldsymbol{B}_1^l = \begin{bmatrix} 1 & 0 & 0 \\ 0 & 3 & 0 \\ 0 & 0 & 2 \end{bmatrix},\ \boldsymbol{B}_2^l = \begin{bmatrix} 1 & 1 & 0 \\ 0 & 2 & 1 \end{bmatrix}$$

Since the three rows of $\boldsymbol{B}_1^l$ are linear independent and the two rows of $\boldsymbol{B}_2^l$ are linear independent too, the system is controllable.

3.2.4 Transformation into Controllable Canonical Form

If a single-input system is controllable, then it can be transformed into the controllable canonical form.

Proof Consider n-dimensional single-input system

$$\dot{\boldsymbol{x}}(t) = \boldsymbol{A}\boldsymbol{x}(t) + \boldsymbol{b}u(t)$$

The system is controllable by assumption, hence the controllability matrix

$$\boldsymbol{Q}_\mathrm{c} = [\boldsymbol{b} \quad \boldsymbol{A}\boldsymbol{b} \quad \cdots \quad \boldsymbol{A}^{n-1}\boldsymbol{b}]$$

has full rank. Assume that the characteristic polynomial of the matrix $\boldsymbol{A}$ is

$$\Delta(\lambda) = |\lambda\boldsymbol{I} - \boldsymbol{A}| = \lambda^n + a_1\lambda^{n-1} + \cdots + a_{n-1}\lambda + a_n \tag{3.10}$$

Let $\boldsymbol{x} = \boldsymbol{P}_\mathrm{c}\boldsymbol{z}$, where the transformation matrix is

$$\boldsymbol{P}_\mathrm{c} = [\boldsymbol{b} \quad \boldsymbol{A}\boldsymbol{b} \quad \cdots \quad \boldsymbol{A}^{n-1}\boldsymbol{b}]\begin{bmatrix} a_{n-1} & a_{n-2} & \cdots & a_1 & 1 \\ a_{n-2} & a_{n-3} & \cdots & 1 & 0 \\ \vdots & \vdots & & \vdots & \vdots \\ a_1 & 1 & \cdots & 0 & 0 \\ 1 & 0 & \cdots & 0 & 0 \end{bmatrix} \tag{3.11}$$

Denoting

$$P_c = [p_{c1} \quad p_{c2} \quad \cdots \quad p_{cn}]$$

we have

$$\begin{aligned}
p_{cn} &= b \\
p_{c,n-1} &= (A + a_1 I) b \\
p_{c,n-2} &= (A^2 + a_1 A + a_2 I) b \\
&\vdots \\
p_{c1} &= (A^{n-1} + a_1 A^{n-2} + \cdots + a_{n-1} I) b
\end{aligned}$$

Observe that

$$\begin{aligned}
Ap_{c1} = {} & Aa_{n-1}b + a_{n-2}A^2 b + \cdots + a_1 A^{n-1} b + A^n b = \\
& (a_n I + a_{n-1}A + \cdots + a_1 A^{n-1} + A^n) b - a_n b = -a_n b = -a_n p_{cn} = \\
& [p_{c1} \quad p_{c2} \quad \cdots \quad p_{cn}][0 \quad \cdots \quad 0 \quad -a_n]^{\mathrm{T}} \\
Ap_{c2} = {} & Aa_{n-2}b + a_{n-3}A^2 b + \cdots + a_1 A^{n-2} b + A^{n-1} b = \\
& A^{-1}(a_{n-2}A^2 + a_{n-3}A^3 + \cdots + a_1 A^{n-1} + A^n) b = A^{-1}(-a_n I - a_{n-1} A) b = \\
& A^{-1}(-a_{n-1}A - a_n I) p_{cn} = -a_{n-1} p_{cn} - a_n A^{-1} p_{cn} = -a_{n-1} p_{cn} + p_{c1} = \\
& -a_n A^{-1} p_{cn} - a_{n-1} p_{cn} = p_{c1} - a_{n-1} p_{cn} = \\
& [p_{c1} \quad p_{c2} \quad \cdots \quad p_{cn}][1 \quad 0 \quad \cdots \quad 0 \quad -a_{n-1}]^{\mathrm{T}} \\
Ap_{ci} = {} & p_{c,i-1} - a_{n-i+1} p_{cn} \\
Ap_{cn} = {} & p_{c,n-1} - a_1 p_{cn} = [p_{c1} \quad p_{c2} \quad \cdots \quad p_{cn}][0 \quad \cdots \quad 0 \quad 1 \quad -a_n]^{\mathrm{T}}
\end{aligned}$$

Thus we have

$$P_c^{-1} A P_c = P_c^{-1} [p_{c1} \quad p_{c2} \quad \cdots \quad p_{cn}] \begin{bmatrix} 0 & 1 & \cdots & 0 & 0 \\ 0 & 0 & \cdots & 0 & 0 \\ \vdots & \vdots & & \vdots & \vdots \\ 0 & 0 & \cdots & 0 & 1 \\ -a_n & -a_{n-1} & \cdots & -a_2 & -a_1 \end{bmatrix} =$$

$$\begin{bmatrix} 0 & 1 & \cdots & 0 & 0 \\ 0 & 0 & \cdots & 0 & 0 \\ \vdots & \vdots & & \vdots & \vdots \\ 0 & 0 & \cdots & 0 & 1 \\ -a_n & -a_{n-1} & \cdots & -a_2 & -a_1 \end{bmatrix}$$

$$P_c^{-1} B = P_c^{-1} p_{cn} = P_c^{-1} [p_{c1} \quad p_{c2} \quad \cdots \quad p_{cn}] \begin{bmatrix} 0 \\ \vdots \\ 0 \\ 1 \end{bmatrix} = \begin{bmatrix} 0 \\ \vdots \\ 0 \\ 1 \end{bmatrix}$$

Example 3.6 Transform the following controllable single-variable state-space representation

$$\dot{x} = \begin{bmatrix} 1 & 0 \\ -1 & 2 \end{bmatrix} x + \begin{bmatrix} -1 \\ 1 \end{bmatrix} u$$

$$y = [0 \quad 1] x$$

into the controllable canonical form.

Solution The characteristic polynomial of the given matrix A is

$$\Delta(\lambda) = |\lambda I - A| = \lambda^2 - 3\lambda + 2$$

hence $a_1 = -3$ and $a_2 = 2$. The controllability matrix Q_c is

$$Q_c = \begin{bmatrix} -1 & -1 \\ 1 & 3 \end{bmatrix}$$

From Eq. (3.10) and Eq. (3.11) we have

$$P_c = \begin{bmatrix} -1 & -1 \\ 1 & 3 \end{bmatrix} \begin{bmatrix} -3 & 1 \\ 1 & 0 \end{bmatrix} = \begin{bmatrix} 0.5 & 0.5 \\ 0 & 1 \end{bmatrix}$$

Hence, introducing transformation $x = P_c z$ the resulting controllable canonical form is

$$\dot{z} = \begin{bmatrix} 0 & 1 \\ -2 & 3 \end{bmatrix} z + \begin{bmatrix} 0 \\ 1 \end{bmatrix} u$$

$$y = [0 \quad 1] z$$

This conclusion is still hold for the multiple-input systems, i.e., a controllable state-space representation can be transformed into a controllable canonical form. Unlike the single-input systems, however, the controllable canonical form of the multiple-input systems is not unique.

3.2.5 Output Controllability

In practical engineering design, the output should be controlled. Similar to the state controllability of a system, we may define controllability for the output vector of a system. Consider the n-dimensional linear time-invariant continuous system

$$\left.\begin{aligned} \dot{x}(t) &= Ax(t) + Bu(t) \\ y(t) &= Cx(t) + Du(t) \end{aligned}\right\} \tag{3.12}$$

where A, B, C and D are $n \times n$, $n \times p$, $q \times n$, $q \times p$ real constant matrices, respectively.

A system is said to be output controllable if there exist a finite time $t_f > t_0$ and an unconstrained control $u(t)$ over $[t_0, t_f]$ that will transfer any nonzero initial output $y(t_0)$ to $y(t_f) = 0$.

It can be proven that a system described by Eq. (3.12) is output controllable if and only if the $q \times np$ matrix $[CB \quad CAB \quad \cdots \quad CA^{n-1}B \quad D]$ has full rank, i.e.,

$$\text{rank}[CB \quad CAB \quad \cdots \quad CA^{n-1}B \quad D] = q \tag{3.13}$$

The matrix in Eq (3.13) is called output controllability matrix.

Example 3.7 Determine the output controllability and state controllability if the system coefficient matrices are

$$A = \begin{bmatrix} -4 & 5 \\ 1 & 0 \end{bmatrix}, \ B = \begin{bmatrix} -5 \\ 1 \end{bmatrix}, \ C = [1 \ \ 1], \ D = 1$$

Solution Calculating the rank of the output controllability matrix yields

$$\text{rank}[\boldsymbol{CB} \ \ \boldsymbol{CAB} \ \ \boldsymbol{D}] = \text{rank}[-6 \ \ 30 \ \ 1] = 1$$

It is equal to q, hence the system is output controllable. Then, calculating the rank of the controllable matrix yields

$$\text{rank}[\boldsymbol{B} \ \ \boldsymbol{AB}] = \text{rank}\begin{bmatrix} 5 & 25 \\ 1 & -5 \end{bmatrix} = 1$$

Obviously, $\text{rank}\boldsymbol{Q}_c < n = 2$, hence the system is not (completely state) controllable.

The state controllability is defined for the state equation, whereas the output controllability is defined for the input-output description; therefore, there is no necessary relation between these two concepts.

3.3 Observability of Linear Continuous Systems

3.3.1 Definition of Observability

Roughly speaking, the problem of observability is to study the possibility of estimating states from the output. If a system is observable, all the response modes of the system can be observed at the output.

For a n-dimensional linear time-invariant system described by

$$\left.\begin{aligned} \dot{\boldsymbol{x}}(t) &= \boldsymbol{Ax}(t) + \boldsymbol{Bu}(t) \\ \boldsymbol{y}(t) &= \boldsymbol{Cx}(t) + \boldsymbol{Du}(t) \end{aligned}\right\} \tag{3.14}$$

the system is said to be (completely state) observable if there exists a finite time $t_1 > t_0$ such that any initial state $\boldsymbol{x}(t_0)$ can be determined from the measurements of the output $\boldsymbol{y}(t)$ over the time interval $[t_0, t_1]$. Without lose of generality, we can assume that $t_0 = 0$.

From Eq. (3.14) we have

$$\boldsymbol{x}(t) = \mathrm{e}^{\boldsymbol{A}t}\boldsymbol{x}(0) + \int_0^t \mathrm{e}^{\boldsymbol{A}(t-\tau)}\boldsymbol{Bu}(\tau)\mathrm{d}\tau$$

and

$$\boldsymbol{y}(t) = \boldsymbol{C}\mathrm{e}^{\boldsymbol{A}t}\boldsymbol{x}(0) + \boldsymbol{C}\int_0^t \mathrm{e}^{\boldsymbol{A}(t-\tau)}\boldsymbol{Bu}(\tau)\mathrm{d}\tau + \boldsymbol{Du}(t) \tag{3.15}$$

Since the concept of observability is defined under the assumption that the matrices $\boldsymbol{A}$, $\boldsymbol{B}$, $\boldsymbol{C}$, $\boldsymbol{D}$ and the input are known, hence the last two terms on the right side of above equation are known and can be separated from the total output in the study of observability. Or, for convenience, we usually choose $\boldsymbol{u}(t) \equiv \boldsymbol{0}$ in Eq. (3.15).

3.3.2 Rank Criterion of Observability

We now consider the n-dimensional linear time-invariant continuous system

$$\left.\begin{aligned} \dot{x}(t) &= Ax(t) \\ y(t) &= Cx(t) \end{aligned}\right\} \tag{3.16}$$

The following rank criterion of observability gives one method of examining observability.

For the system described by Eq. (3.16) to be observable if and only if the matrix

$$Q_o = \begin{bmatrix} C \\ CA \\ \vdots \\ CA^{n-1} \end{bmatrix} \tag{3.17}$$

has rank n. The matrix Q_o is described in terms of A and C, and is called the observability matrix.

Proof From Eq. (3.15) the output for the system given by Eq. (3.16) is

$$y(t) = Ce^{At}x(0)$$

Making use of Eq. (3.4) yields

$$y(t) = C\sum_{i=0}^{n-1} \alpha_i(t)A^i x(0) = \sum_{i=0}^{n-1} \alpha_i(t)CA^i x(0)$$

or

$$y(t) = [\alpha_0 I \quad \alpha_1 I \quad \cdots \quad \alpha_{n-1} I]\begin{bmatrix} C \\ CA \\ \vdots \\ CA^{n-1} \end{bmatrix} \tag{3.18}$$

Therefore the system is observable if and only if there exists a set of n linear independent row vectors in Q_o, i.e., the observability matrix has rank n. For a system with one output, the matrix Q_o is square, thus the condition of observability is that Q_o must be nonsingular.

Since only the matrices A and C are involved for examining the observability of a system, sometimes we say that a pair (A, C) is observable, which means that Q_o is of rank n.

Example 3.8 Examine the observability of the system described by

$$\dot{x} = \begin{bmatrix} 0 & 1 & 0 \\ 0 & 0 & 1 \\ -6 & -11 & -6 \end{bmatrix} x + \begin{bmatrix} 0 \\ 0 \\ 1 \end{bmatrix} u$$

$$y = [4 \quad 5 \quad 1]x.$$

Solution From the given state-space representation we get

$$CA = [4 \quad 5 \quad 1]\begin{bmatrix} 0 & 1 & 0 \\ 0 & 0 & 1 \\ -6 & -11 & -6 \end{bmatrix} = [-6 \quad -7 \quad -1]$$

$$CA^2 = [-6 \quad -7 \quad -1]\begin{bmatrix} 0 & 1 & 0 \\ 0 & 0 & 1 \\ -6 & -11 & -6 \end{bmatrix} = [6 \quad 5 \quad -1]$$

Thus, the observability matrix is

$$Q_o = \begin{bmatrix} C \\ CA \\ CA^2 \end{bmatrix} = \begin{bmatrix} 4 & 5 & 1 \\ -6 & -7 & -1 \\ 6 & 5 & -1 \end{bmatrix}$$

Since

$$\text{rank } Q_o = 2 < n = 3$$

the system is unobservable.

Example 3.9 Prove that a single-output system with the state-space representation in observable canonical form must be observable.

Proof The state-apace representation can be represented as

$$\dot{x} = \begin{bmatrix} 0 & \cdots & 0 & * \\ 1 & \cdots & 0 & * \\ \vdots & & \vdots & \vdots \\ 0 & \cdots & 1 & * \end{bmatrix} x, \ y = [0 \ \cdots \ 0 \ 1]$$

where * s denote possible nonzero elements. The observability matrix is

$$Q_o = \begin{bmatrix} 0 & 0 & \cdots & 1 \\ \vdots & \vdots & & \vdots \\ 0 & 1 & \cdots & * \\ 1 & * & \cdots & * \end{bmatrix}$$

Q_o must be nonsingular because it is a triangular matrix. Hence, the system is always observable if it is described by the observable canonical form.

3.3.3 Observability Criterion for Jordan Canonical Form

1. All Eigenvalues Are Distinct

If all the eigenvalues of a system matrix are distinct, assuming that $u = 0$, *the derived Jordan canonical form is*

$$\dot{x} = \begin{bmatrix} \lambda_1 & & & \\ & \lambda_2 & & \\ & & \ddots & \\ & & & \lambda_n \end{bmatrix} x$$

$$y = [c_1 \ \ c_2 \ \ \cdots \ \ c_n] x$$

In this case, the matrix pair (A, C) is observable if and only if there is not any column with all zero components in the matrix C.

For example, the system

$$\dot{x} = \begin{bmatrix} -2 & 0 \\ 0 & -3 \end{bmatrix} x$$

$$y = [1 \ \ 2] x$$

is observable; and the system

$$\dot{x} = \begin{bmatrix} -2 & 0 \\ 0 & -3 \end{bmatrix} x$$
$$y = [1 \quad 0] x$$

is unobservable.

2. Not All Eigenvalues Are Distinct

Assume that the matrices A, B and C of derived Jordan form are as shown in Eq. (3.8). In this case, the system is observable if and only if, for each $i = 1, 2, \cdots, m$, all the columns of the $q \times \alpha_i$ matrix

$$C_i^1 = [c_{1i1} \quad c_{1i2} \quad \cdots \quad c_{1i\alpha_i}] \tag{3.19}$$

are linearly independent.

Example 3.10 Consider the observability of the following systems.

(1) $\dot{x} = \begin{bmatrix} -2 & 0 \\ 0 & -2 \end{bmatrix} x$;
$y = [1 \quad 2] x$

(2) $\dot{x} = \left[\begin{array}{ccc:cc} -2 & & & & \\ & -2 & 1 & & \\ & & -2 & & \\ \hdashline & & & -5 & 1 \\ & & & & -5 \end{array}\right] x$;

$$y = \begin{bmatrix} 1 & 2 & 0 & 2 & 2 \\ 0 & 0 & 1 & 0 & 1 \end{bmatrix} x$$

(3) $\dot{x} = \left[\begin{array}{cccc:ccc} 2 & 1 & & & & & \\ & 2 & & & & & \\ & & 2 & & & & \\ & & & 2 & & & \\ \hdashline & & & & 3 & 1 & \\ & & & & & 3 & \\ & & & & & & 3 \end{array}\right] x$

$$y = \begin{bmatrix} 1 & 0 & 0 & 0 & 1 & 0 & 0 \\ 0 & 0 & 1 & 0 & 1 & 0 & 1 \\ 0 & 0 & 0 & 2 & 0 & 0 & 0 \end{bmatrix} x.$$

Solution (1) There is only one distinct eigenvalue and $\alpha_1 = 2$. Since

$$C_1^l = [1 \quad 2]$$

in this matrix the two constants are always linear dependent, hence the system is unobservable.

(2) Since

$$B_1^l = \begin{bmatrix} 1 & 2 \\ 0 & 0 \end{bmatrix}$$

where the two columns are linear dependent, thus the system is unobservable.

(3) From Eq. (3.19) we have

$$C_1^1 = \begin{bmatrix} 1 & 0 & 0 \\ 0 & 1 & 0 \\ 0 & 0 & 2 \end{bmatrix},\ C_2^1 = \begin{bmatrix} 1 & 0 \\ 1 & 1 \\ 0 & 0 \end{bmatrix}$$

Since the three columns of C_1^1 are linear independent and the two columns of C_2^1 are linear independent too, hence the system is observable.

3.3.4 Transformation into Observability Canonical Form

Similar to the case of controllability, it can be proven that if the n-dimensional single-output system

$$\dot{x}(t) = Ax(t) + BU(t)$$
$$y(t) = Cx(t)$$

is observable, then it can be transformed into the observable canonical form by introducing a nonsingular transformation $x = P_o z$. The transformation matrix P_o may be given from

$$P_o^{-1} = \begin{bmatrix} a_{n-1} & a_{n-2} & \cdots & a_1 & 1 \\ a_{n-2} & a_{n-3} & \cdots & 1 & 0 \\ \vdots & \vdots & & \vdots & \vdots \\ a_1 & 1 & \cdots & 0 & 0 \\ 1 & 0 & \cdots & 0 & 0 \end{bmatrix} Q_o \tag{3.20}$$

where $a_1, a_2, \cdots, a_n$ are the coefficients of the characteristic polynomial of the matrix A, and Q_o is the observability matrix of the system.

Example 3.11 Transform the following observable single-variable state-space representation

$$\dot{x} = \begin{bmatrix} 1 & 0 \\ -1 & 2 \end{bmatrix} x + \begin{bmatrix} -1 \\ 1 \end{bmatrix} u$$
$$y = \begin{bmatrix} 0 & 1 \end{bmatrix} x$$

into the observable canonical form.

Solution Since the observability matrix

$$Q_o = \begin{bmatrix} 0 & 1 \\ -1 & 2 \end{bmatrix}$$

has full rank, the system is observable. The characteristic polynomial of the given matrix A is

$$\Delta(\lambda) = |\lambda I - A| = \lambda^2 - 3\lambda + 2$$

Letting $x = P_o z$ and choosing

$$P_o^{-1} = \begin{bmatrix} -3 & 1 \\ 1 & 0 \end{bmatrix} \begin{bmatrix} 0 & 1 \\ -1 & 2 \end{bmatrix} = \begin{bmatrix} -1 & -1 \\ 0 & 1 \end{bmatrix},\ P_o = \begin{bmatrix} -1 & -1 \\ 0 & 1 \end{bmatrix}$$

yield the observable canonical form

$$\dot{z} = \begin{bmatrix} 0 & -2 \\ 1 & 3 \end{bmatrix} z + \begin{bmatrix} 0 \\ 1 \end{bmatrix} u$$
$$y = \begin{bmatrix} 0 & 1 \end{bmatrix} x$$

Unlike the single-output systems, the observable canonical form of a multiple-output system is not unique.

3.4 Controllability and Observability of Discrete -Time Systems

In the previous sections, we have studied the controllability and observability of continuous-time systems. Very similar analyses and results can be obtained for discrete-time systems, where it is assumed that the inputs and outputs are known (or are of interest) only at discrete time instants and these instants are all integral multiples of some basic unit T, say

$$t_0 = 0,\ t_1 = T,\ t_2 = 2T,\ \cdots$$

Also, it is assumed that the time parameter, denoted by k, takes integral values, $k = 0, \pm 1, \pm 2, \cdots$

The discrete-time state-space representation is usually written in the form

$$\left.\begin{aligned} \boldsymbol{x}(k+1) &= \boldsymbol{A}\boldsymbol{x}(k) + \boldsymbol{B}\boldsymbol{u}(k) \\ \boldsymbol{y}(k) &= \boldsymbol{C}\boldsymbol{x}(k) \end{aligned}\right\} \tag{3.21}$$

3.4.1 Controllability of Linear Discrete-time Systems

The discrete-time system of Eq. (3.21) is said to be controllable at time instant l, if for any nonzero state $\boldsymbol{x}(l)$ there exist a finite instant $m > l$ and a corresponding input $\boldsymbol{u}(k)$, at the time instants over $[l, m]$, that will transfer $\boldsymbol{x}(l)$ to the state $\boldsymbol{x}(m) = \boldsymbol{0}$.

If we assume that the dimension of a system is n and define a controllability matrix $\boldsymbol{Q}_c$ as

$$\boldsymbol{Q}_c = [\boldsymbol{B} \quad \boldsymbol{AB} \quad \cdots \quad \boldsymbol{A}^{n-1}\boldsymbol{B}] \tag{3.22}$$

then a rank criterion of controllability for the discrete-time system, similar to that for the continuous systems, can be obtained. The system described by Eq. (3.21) is controllable if and only if

$$\operatorname{rank} \boldsymbol{Q}_o = n \tag{3.23}$$

3.4.2 Observability of Linear Discrete-time Systems

The discrete-time system of Eq. (3.21) is said to be observable at time instant l if there exists a finite instant $m > l$ such that any nonzero state $\boldsymbol{x}(l)$ can be uniquely determined from the output $\boldsymbol{y}(k)$ at the time instants over $[l, m]$.

The n-dimensional discrete-time system described by Eq. (3.21) is observable if and only if the observability matrix has rank n. The observability matrix is define as

$$\boldsymbol{Q}_o = \begin{bmatrix} \boldsymbol{C} \\ \boldsymbol{CA} \\ \vdots \\ \boldsymbol{CA}^{n-1} \end{bmatrix} \tag{3.24}$$

3.4.3 Controllability and Observability of Discretized Systems

The following example will show that the controllability and observability of a discretized system may be changed.

Example 3.12 Analyze controllability and observability of the discretized system if the original continuous-time system is given by

$$\dot{x} = \begin{bmatrix} 0 & \omega \\ -\omega & 0 \end{bmatrix} x + \begin{bmatrix} 0 \\ 1 \end{bmatrix} u, \omega > 0$$

$$y = [1 \quad 0] x$$

and the sampling period is T.

Solution Before discretization, since

$$\text{rank } Q_c = \text{rank} \begin{bmatrix} 0 & \omega \\ 1 & 0 \end{bmatrix} = 2 = n$$

$$\text{rank } Q_o = \text{rank} \begin{bmatrix} 1 & 0 \\ 0 & \omega \end{bmatrix} = 2 = n$$

the continuous-time system is controllable and observable.

Using Laplace transform, the matrix exponential function is

$$e^{At} = \mathscr{L}^{-1} \begin{bmatrix} s & -1 \\ \omega^2 & s \end{bmatrix}^{-1} = \mathscr{L}^{-1} \begin{bmatrix} \dfrac{s}{s^2 + \omega^2} & \dfrac{1}{s^2 + \omega^2} \\ \dfrac{-\omega^2}{s^2 + \omega^2} & \dfrac{s}{s^2 + \omega^2} \end{bmatrix} = \begin{bmatrix} \cos \omega t & \dfrac{1}{\omega} \sin \omega t \\ -\omega \sin \omega t & \cos \omega t \end{bmatrix}$$

Then the system matrix and input matrix of the discretized system are

$$A^*(T) = e^{AT} = \begin{bmatrix} \cos \omega T & \dfrac{1}{\omega} \sin \omega T \\ -\omega \sin \omega T & \cos \omega T \end{bmatrix}$$

$$B^*(T) = \int_0^T e^{AT} B \, dt = \int_0^T \begin{bmatrix} \cos \omega t & \dfrac{1}{\omega} \sin \omega t \\ -\omega \sin \omega t & \cos \omega t \end{bmatrix} \begin{bmatrix} 0 \\ 1 \end{bmatrix} dt = \begin{bmatrix} \dfrac{1 - \cos \omega T}{\omega^2} \\ \dfrac{\sin \omega t}{\omega} \end{bmatrix}$$

and the discretized state-space representation is

$$x(k+1) = \begin{bmatrix} \cos \omega T & \dfrac{1}{\omega} \sin \omega T \\ -\omega \sin \omega T & \cos \omega T \end{bmatrix} x(k) + \begin{bmatrix} \dfrac{1 - \cos \omega T}{\omega^2} \\ \dfrac{\sin \omega t}{\omega} \end{bmatrix} u(k)$$

$$y(k) = [1 \quad 0] x(k)$$

The controllability matrix is

$$Q_c = \begin{bmatrix} \dfrac{1 - \cos \omega T}{\omega^2} & \dfrac{\cos \omega T - \cos^2 \omega t + \sin^2 \omega t}{\omega^2} \\ \dfrac{\sin \omega t}{\omega} & \dfrac{2 \sin \omega T \cos \omega T - \sin \omega t}{\omega} \end{bmatrix} = \begin{bmatrix} 1 - \cos \omega T & \cos \omega T - \cos 2\omega T \\ \sin \omega T & -\sin \omega T + \sin 2\omega T \end{bmatrix}$$

It can be seen that the discretized system is controllable if and only if $T \neq \frac{k\pi}{\omega}$, $k = 0, \pm 1, \pm 2, \cdots$

The observability matrix of the discretized system is

$$\boldsymbol{Q}_o = \begin{bmatrix} 1 & 0 \\ \cos \omega T & \frac{\sin \omega T}{\omega} \end{bmatrix}$$

Obviously, the discretized system is observable if and only if $T \neq \frac{k\pi}{\omega}$, $k = 0, \pm 1, \pm 2, \cdots$

3.5 Controllability, Observability and Transfer Function

Although controllability and observability are concepts of modern control theory, they are closely related to the properties of the transfer function.

If the transfer function of a single-variable system has pole-zero cancellation, the system will be either uncontrollable or unobservable, depending on how the state variables are chosen. If the transfer function of a single variable system does not have pole-zero cancellation, the system can always be described by state-space representation as a controllable and observable system.

Proof Consider that a single-variable system with distinct eigenvalues is described by

$$\left.\begin{aligned} \dot{\boldsymbol{x}}(t) &= \boldsymbol{A}\boldsymbol{x}(t) + \boldsymbol{b}u(t) \\ y &= \boldsymbol{c}\boldsymbol{x}(t) \end{aligned}\right\} \tag{3.25}$$

Diagonalizing the matrix $\boldsymbol{A}$, the new state-space description in canonical form is

$$\left.\begin{aligned} \dot{\boldsymbol{x}} &= \begin{bmatrix} \lambda_1 & & & \\ & \lambda_2 & & \\ & & \ddots & \\ & & & \lambda_n \end{bmatrix} \boldsymbol{x} + \begin{bmatrix} \beta_1 \\ \beta_2 \\ \vdots \\ \beta_n \end{bmatrix} u \\ y &= [\gamma_1 \quad \gamma_2 \quad \cdots \quad \gamma_n] \boldsymbol{x} \end{aligned}\right\} \tag{3.26}$$

Taking Laplace transform and assuming zero initial conditions we have the transfer function between the output and input as

$$\frac{Y(s)}{U(s)} = [\gamma_1 \quad \gamma_2 \quad \cdots \quad \gamma_n] \begin{bmatrix} s - \lambda_1 & & & \\ & s - \lambda_2 & & \\ & & \ddots & \\ & & & s - \lambda_n \end{bmatrix}^{-1} \begin{bmatrix} \beta_1 \\ \beta_2 \\ \vdots \\ \beta_n \end{bmatrix} =$$

$$[\gamma_1 \quad \gamma_2 \quad \cdots \quad \gamma_n]\begin{bmatrix} \frac{1}{s-\lambda_1} & & & \\ & \frac{1}{s-\lambda_2} & & \\ & & \ddots & \\ & & & \frac{1}{s-\lambda_n} \end{bmatrix}\begin{bmatrix} \beta_1 \\ \beta_2 \\ \vdots \\ \beta_n \end{bmatrix} =$$

$$\frac{\gamma_1\beta_1}{s-\lambda_1} + \frac{\gamma_2\beta_2}{s-\lambda_2} + \cdots + \frac{\gamma_n\beta_n}{s-\lambda_n} = \sum_{i=1}^{n}\frac{\gamma_i\beta_i}{s-\lambda_i} \tag{3.27}$$

As we know, the system described by Eq. (3.25) is controllable and/or observable if and only if $\beta_i \neq 0$ and/or $\gamma_i \neq 0$ for $i = 1,2,\cdots,n$.

Since all the eigenvalues of matrix $\boldsymbol{A}$ are distinct, the transfer function can be rewritten in the zero-pole form as

$$\frac{Y(s)}{U(s)} = \frac{k(s-z_1)(s-z_2)\cdots(s-z_m)}{(s-\lambda_1)(s-\lambda_2)\cdots(s-\lambda_n)},\ n > m \tag{3.28}$$

which can be expanded by partial fraction into

$$\frac{Y(s)}{U(s)} = \sum_{i=1}^{n}\frac{\delta_i}{s-\lambda_i},\ i = 1,2,\cdots,n \tag{3.29}$$

where $\delta_i = \lim\limits_{s\to\lambda_i}(s-\lambda_i)\frac{Y(s)}{U(s)},\ i = 1,2,\cdots,n$. If $\frac{Y(s)}{U(s)}$ has one or more pairs of identical pole-zero, for instance $\lambda_1 = z_1$ in Eq. (3.28), then $\delta_1 = 0$ in Eq. (3.29).

Comparing Eq. (3.27) and Eq. (3.29) yields

$$\delta_i = \gamma_i\beta_i,\ i = 1,2,\cdots,n \tag{3.30}$$

Therefore, when $\delta_i = 0$, β_i and/or γ_i will be zero, and the state x_i is uncontrollable and/or unobservable, i.e. the system is uncontrollable and/or unobservable. When $\delta_i \neq 0$, both β_i and γ_i are from zero, and the system is controllable and observable.

Example 3.13 Consider the controllatility and observatility of the system with transfer fumction

$$\frac{Y(s)}{U(s)} = \frac{s+2.5}{s^2+1.5s-2.5}$$

Solution Since

$$\frac{Y(s)}{U(s)} = \frac{s+2.5}{(s+2.5)(s-1)} = \frac{0}{s+2.5} + \frac{1}{s-1}$$

which has an identical pole and zero at $s = -2.5$, we can write its state-space representation in three canonical forms

(1) $\dot{\boldsymbol{x}} = \begin{bmatrix} 0 & 1 \\ 2.5 & -1.5 \end{bmatrix}\boldsymbol{x} + \begin{bmatrix} 0 \\ 1 \end{bmatrix}u,\ y = [1 \quad 0]\boldsymbol{x}$

(2) $\dot{\boldsymbol{x}} = \begin{bmatrix} 0 & 2.5 \\ 1 & -1.5 \end{bmatrix}\boldsymbol{x} + \begin{bmatrix} 1 \\ 0 \end{bmatrix}u,\ y = [0 \quad 1]\boldsymbol{x}$

(3) $\dot{x} = \begin{bmatrix} -2.5 & 0 \\ 0 & 1 \end{bmatrix} x + \begin{bmatrix} 0 \\ 1 \end{bmatrix} u, \ y = [0 \quad 1] x$

In case (1), for the controllable canonical form, the system is controllable but not observable because

$$\operatorname{rank} Q_o = \operatorname{rank}\begin{bmatrix} 1 & 0 \\ 1 & 0 \end{bmatrix} = 1 < n$$

In case (2), for the observable canonical form, the system is observable but not controllable because

$$\operatorname{rank} Q_c = \operatorname{rank}\begin{bmatrix} 1 & 1 \\ 0 & 0 \end{bmatrix} = 1 < n$$

In case (3), for the Jordan canonical form, it is easy to get the conclusion that the system is neither controllable nor observable.

3.6 Principle of Duality

It can be seen from previous discussion that the controllability and observability are conceptually and formally similar. This phenomenon is determined by the principle of duality of system.

3.6.1 Dual System

Two linear systems (A_1, B_1, C_1) and (A_2, B_2, C_2) are said to be dual if $A_1^T = A_2$, $B_1^T = C_2$, and $C_1^T = B_2$. That is, if a system is described by

$$\left.\begin{aligned} \dot{x}(t) &= Ax(t) + Bu(t) \\ y &= Cx(t) \end{aligned}\right\} \tag{3.31}$$

then its dual system is defined as

$$\left.\begin{aligned} \dot{z} &= A^T z + C^T v \\ w &= B^T z \end{aligned}\right\} \tag{3.32}$$

where x and z are $n \times 1$ state vectors; u and v are $p \times 1$ and $q \times 1$ input vectors, respectively; y and w are $q \times 1$ and $p \times 1$ output vectors, respectively; A is a $n \times n$ matrix, B is a $n \times p$ matrix, C is a $q \times n$ matrix; and A^T, B^T, C^T are the transpose matrices of A, B, C, respectively.

3.6.2 Principle of Duality

Denote Q_{ci} and Q_{oi} be the controllability matrix and observability matrix of the ith, i = 1,2, system. Writing the rank conditions of controllability and observability to the dual systems, respectively, we have

$$\begin{aligned} \operatorname{rank} Q_{o2} &= \operatorname{rank}[C_2^T \quad (C_2A_2)^T \quad \cdots \quad (C_2A_2^{n-1})^T] = \\ &\operatorname{rank}[C_2^T \quad A_2^T C_2^T \quad \cdots \quad (A_2^{n-1})^T C_2^T] = \end{aligned}$$

$$\begin{aligned}
&\mathrm{rank}[(\boldsymbol{B}_1^{\mathrm{T}})^{\mathrm{T}} \quad (\boldsymbol{A}_1^{\mathrm{T}})^{\mathrm{T}}(\boldsymbol{B}_1^{\mathrm{T}})^{\mathrm{T}} \quad \cdots \quad ((\boldsymbol{A}_1^{\mathrm{T}})^{n-1})^{\mathrm{T}}(\boldsymbol{B}_1^{\mathrm{T}})^{\mathrm{T}}] = \\
&\mathrm{rank}[\boldsymbol{B}_1 \quad \boldsymbol{A}_1\boldsymbol{B}_1 \quad \cdots \quad \boldsymbol{A}_1^{n-1}\boldsymbol{B}_1] = \\
&\mathrm{rank}\boldsymbol{Q}_{c1}
\end{aligned}$$

$$\begin{aligned}
\mathrm{rank}\boldsymbol{Q}_{c2} &= \mathrm{rank}[\boldsymbol{B}_2 \quad \boldsymbol{A}_2\boldsymbol{B}_2 \quad \cdots \quad \boldsymbol{A}_2^{n-1}\boldsymbol{B}_2] = \\
&\mathrm{rank}[\boldsymbol{C}_1^{\mathrm{T}} \quad \boldsymbol{A}_1^{\mathrm{T}}\boldsymbol{C}_1^{\mathrm{T}} \quad \cdots \quad (\boldsymbol{A}_1^{\mathrm{T}})^{n-1}\boldsymbol{C}_1^{\mathrm{T}}] = \\
&\mathrm{rank}[\boldsymbol{C}_1^{\mathrm{T}} \quad (\boldsymbol{C}_1\boldsymbol{A}_1)^{\mathrm{T}} \quad \cdots \quad (\boldsymbol{C}_1\boldsymbol{A}_1^{n-1})^{\mathrm{T}}] = \\
&\mathrm{rank}\boldsymbol{Q}_{o1}
\end{aligned}$$

From above relationships it is not difficult to get the following principle of duality: If two n-dimension linear systems $(\boldsymbol{A}_1, \boldsymbol{B}_1, \boldsymbol{C}_1)$ and $(\boldsymbol{A}_2, \boldsymbol{B}_2, \boldsymbol{C}_2)$ are dual systems, then

(a) the system $(\boldsymbol{A}_1, \boldsymbol{B}_1, \boldsymbol{C}_1)$ is controllable if and only if the system $(\boldsymbol{A}_2, \boldsymbol{B}_2, \boldsymbol{C}_2)$ is observable;

(b) the system $(\boldsymbol{A}_1, \boldsymbol{B}_1, \boldsymbol{C}_1)$ is observable if and only if the system $(\boldsymbol{A}_2, \boldsymbol{B}_2, \boldsymbol{C}_2)$ is controllable.

3.7 Structural Decomposition

If a system $(\boldsymbol{A}, \boldsymbol{B}, \boldsymbol{C})$ is uncontrollable then usually some of the state variables are controllable and others are uncontrollable. Similarly, if a system$(\boldsymbol{A}, \boldsymbol{B}, \boldsymbol{C})$ is unobservable then usually some of the state variables are observable and others are unobservable. From the point of controllability and observability, the state variables $\boldsymbol{x}$ can be decomposed into four groups: controllable and observable $\boldsymbol{x}_{co}$, controllable but not observable $\boldsymbol{x}_{c\bar{o}}$, uncontrollable but observable $\boldsymbol{x}_{\bar{c}o}$, and neither controllable nor observable $\boldsymbol{x}_{\overline{co}}$, where the subscript c is used to stand for controllable, $\bar{c}$ for uncontrollable, o for observable, and $\bar{o}$ for unobservable. Correspondingly, the system can be decomposed into four subsystems, which is called the structural decomposition.

For a state-space representation in the Jordan canonical form, the system can be decomposed by inspection.

Example 3.14 Decompose the following state-space representation

$$\dot{\boldsymbol{x}} = \begin{bmatrix} \lambda_1 & 1 & & & \\ & \lambda_1 & 1 & & \\ & & \lambda_1 & & \\ & & & \lambda_2 & 1 \\ & & & & \lambda_2 \end{bmatrix}\boldsymbol{x} + \begin{bmatrix} 0 \\ 1 \\ 0 \\ 0 \\ 1 \end{bmatrix}u$$

$$y = [0 \quad 1 \quad 1 \quad 0 \quad 1]\boldsymbol{x}$$

Solution From the pair of matrices $(\boldsymbol{A}, \boldsymbol{B})$ it can be seen that the state variables x_1, x_2, x_4 and x_5 are controllable, x_3 is uncontrollable. From the pair of matrices $(\boldsymbol{A}, \boldsymbol{C})$ it can be seen that the state variables x_2, x_3 and x_5 are observable, x_1 and x_4 are unobservable. Hence, the state variables x_2 and x_5 are controllable and observable, x_1 and x_4 are controllable but unobservable, x_3 is uncontrollable but observable. The state-space representation can be rewritten as

$$\begin{bmatrix}\dot{x}_2\\ \dot{x}_5\\ \dot{x}_1\\ \dot{x}_4\\ \dot{x}_3\end{bmatrix}=\begin{bmatrix}\lambda_1 & 0 & 0 & 0 & 1\\ 0 & \lambda_1 & 0 & 0 & 0\\ 1 & & \lambda_1 & & \\ & 1 & & \lambda_2 & \\ & & & & \lambda_2\end{bmatrix}\begin{bmatrix}x_2\\ x_5\\ x_1\\ x_4\\ x_3\end{bmatrix}+\begin{bmatrix}1\\ 1\\ 0\\ 0\\ 0\end{bmatrix}u,\ y=\begin{bmatrix}1 & 1 & 0 & 0 & 1\end{bmatrix}\begin{bmatrix}x_2\\ x_5\\ x_1\\ x_4\\ x_3\end{bmatrix}$$

In what follows a canonical decomposition method will be described without proof.

3.7.1 Decomposition According to Controllability

Consider a n-dimensional linear time-invariant system $(\boldsymbol{A},\boldsymbol{B},\boldsymbol{C})$. If the controllability matrix has rank $n_c(n_c < n)$, then there exists a nonsingular transformation $\boldsymbol{x} = \boldsymbol{T}_c\boldsymbol{z}$ such that

$$\left.\begin{aligned}\begin{bmatrix}\dot{z}_c\\ \dot{z}_{\bar{c}}\end{bmatrix} &= \boldsymbol{T}_c^{-1}\boldsymbol{A}\boldsymbol{T}_c\begin{bmatrix}z_c\\ z_{\bar{c}}\end{bmatrix}+\boldsymbol{T}_c^{-1}\boldsymbol{B}\boldsymbol{u}=\begin{bmatrix}\bar{\boldsymbol{A}}_c & \bar{\boldsymbol{A}}_{12}\\ \boldsymbol{0} & \bar{\boldsymbol{A}}_{\bar{c}}\end{bmatrix}\begin{bmatrix}z_c\\ z_{\bar{c}}\end{bmatrix}+\begin{bmatrix}\bar{\boldsymbol{B}}_c\\ 0\end{bmatrix}\boldsymbol{u}\\ \begin{bmatrix}y_c\\ y_{\bar{c}}\end{bmatrix} &= \boldsymbol{C}\boldsymbol{T}_c\begin{bmatrix}z_c\\ z_{\bar{c}}\end{bmatrix}=\begin{bmatrix}\bar{\boldsymbol{C}}_c & \bar{\boldsymbol{C}}_{\bar{c}}\end{bmatrix}\begin{bmatrix}z_c\\ z_{\bar{c}}\end{bmatrix}\end{aligned}\right\}\qquad(3.33)$$

where $\boldsymbol{T}_c$ is a nonsingular transformation matrix constructed with any n_c linear independent column vectors of $\boldsymbol{Q}_c$ and $n - n_c$ entirely arbitrary column vectors.

Eq. (3.33) shows that the state variables in $z_{\bar{c}}$ are not affected directly by the input $\boldsymbol{u}$ or indirectly through the state variables in z_c. Hence, the state vector $z_{\bar{c}}$ is uncontrollable. The resulting controllable subsystem is

$$\left.\begin{aligned}\dot{z}_c &= \bar{\boldsymbol{A}}_c z_c + \bar{\boldsymbol{A}}_{12} z_{\bar{c}} + \bar{\boldsymbol{B}}_c\boldsymbol{u}\\ y_c &= \bar{\boldsymbol{C}}_c z_c\end{aligned}\right\}\qquad(3.34)$$

and the uncontrollable subsystem is

$$\left.\begin{aligned}\dot{z}_{\bar{c}} &= \bar{\boldsymbol{A}}_{\bar{c}} z_{\bar{c}}\\ y_{\bar{c}} &= \bar{\boldsymbol{C}}_{\bar{c}} z_{\bar{c}}\end{aligned}\right\}\qquad(3.35)$$

Moreover, the n_c-dimensional subsystem $(\bar{\boldsymbol{A}}_c,\bar{\boldsymbol{B}}_c,\bar{\boldsymbol{C}}_c)$ is controllable and has the same transfer function matrix as $(\boldsymbol{A},\boldsymbol{B},\boldsymbol{C})$.

Example 3.15 Decompose the following state-space representation

$$\dot{\boldsymbol{x}}=\begin{bmatrix}1 & 2 & -1\\ 0 & 1 & 0\\ 1 & -4 & 3\end{bmatrix}\boldsymbol{x}+\begin{bmatrix}0\\ 0\\ 1\end{bmatrix}u,\ y=\begin{bmatrix}1 & -1 & 1\end{bmatrix}\boldsymbol{x}$$

according to the controllability.

Solution The controllability matrix is

$$Q_c=\begin{bmatrix}0 & -1 & -4\\ 0 & 0 & 0\\ 1 & 3 & 8\end{bmatrix}$$

which has rank 2. The transformation matrix can be chosen as

$$T_c = \begin{bmatrix} 0 & -1 & 0 \\ 0 & 0 & 0 \\ 1 & 3 & 1 \end{bmatrix}$$

where the first two columns are the linear independent columns of Q_c and the third column is chosen arbitrarily to make T_c nonsingular. The inverse of T_c is

$$T_c^{-1} = \begin{bmatrix} 3 & 0 & 1 \\ -1 & 0 & 0 \\ 0 & 1 & 0 \end{bmatrix}$$

Letting $x = T_c z$ yields

$$\dot{z} = \begin{bmatrix} \dot{z}_c \\ \dot{z}_{\bar{c}} \end{bmatrix} = \left[\begin{array}{cc:c} 0 & -4 & 2 \\ 1 & 4 & -2 \\ \hdashline 0 & 0 & 1 \end{array}\right] \begin{bmatrix} z_c \\ z_{\bar{c}} \end{bmatrix} + \begin{bmatrix} 1 \\ 0 \\ 0 \end{bmatrix} u$$

$$y = [1 \quad 2 \,\vdots\, -1] \begin{bmatrix} z_c \\ z_{\bar{c}} \end{bmatrix}$$

3.7.2 Decomposition According to Observability

If the observability matrix of the n-dimensional linear time-invariant system (A, B, C) has rank $n_o (n_o < n)$, then there exists a nonsingular transformation $x = T_o z$ such that

$$\left.\begin{aligned} \begin{bmatrix} \dot{z}_o \\ \dot{z}_{\bar{o}} \end{bmatrix} &= T_o^{-1} A T_o \begin{bmatrix} z_o \\ z_{\bar{o}} \end{bmatrix} + T_o^{-1} B u = \begin{bmatrix} \tilde{A}_o & 0 \\ \tilde{A}_{21} & \tilde{A}_{\bar{o}} \end{bmatrix} \begin{bmatrix} z_o \\ z_{\bar{o}} \end{bmatrix} + \begin{bmatrix} \tilde{B}_o \\ \tilde{B}_{\bar{o}} \end{bmatrix} u \\ \begin{bmatrix} y_o \\ y_{\bar{o}} \end{bmatrix} &= C T_o \begin{bmatrix} z_o \\ z_{\bar{o}} \end{bmatrix} = [\tilde{C}_o \quad 0] \begin{bmatrix} z_o \\ z_{\bar{o}} \end{bmatrix} \end{aligned}\right\} \tag{3.36}$$

where T_o^{-1} is a nonsingular matrix and can be constructed with n_o any linear independent rows of Q_o and $n - n_o$ entirely arbitrary row vectors.

Eq. (3.36) shows that the state variables in $z_{\bar{o}}$ don't appear directly in the output y or indirectly through the state variables in z_o. Therefore, the state vector $z_{\bar{o}}$ is unobservable. The resulting observable subsystem is

$$\left.\begin{aligned} \dot{z}_o &= \tilde{A}_o z_o + \tilde{B}_o u \\ y_o &= \tilde{C}_o z_o = y \end{aligned}\right\} \tag{3.37}$$

and the uncontrollable subsystem is

$$\left.\begin{aligned} \dot{z}_{\bar{o}} &= \tilde{A}_{21} z_o + \tilde{A}_{\bar{o}} z_{\bar{c}} + \tilde{B}_{\bar{o}} u \\ y_{\bar{o}} &= 0 \end{aligned}\right\} \tag{3.38}$$

Moreover, the n_o-dimensional subsystem $(\tilde{A}_o, \tilde{B}_o, \tilde{C}_o)$ is observable and has the same transfer function matrix as (A, B, C).

Example 3.16 Decompose the following description

$$\dot{x} = \begin{bmatrix} 1 & 2 & -1 \\ 0 & 1 & 0 \\ 1 & -4 & 3 \end{bmatrix} x + \begin{bmatrix} 0 \\ 0 \\ 1 \end{bmatrix} u,$$

$$y = [1 \quad -1 \quad 1] x$$

according to the observability.

Solution The observability matrix is

$$Q_o = \begin{bmatrix} 1 & -1 & 1 \\ 2 & -3 & 2 \\ 4 & -7 & 4 \end{bmatrix}$$

which has rank 2. The inverse of the transformation matrix can be chosen as

$$T_o^{-1} = \begin{bmatrix} 1 & -1 & 1 \\ 2 & -3 & 2 \\ 0 & 0 & 1 \end{bmatrix}$$

where the first two rows are the linear independent rows of Q_o and the third row is chosen arbitrarily to make T_o^{-1} nonsingular. The transformation matrix is

$$T_o = \begin{bmatrix} 3 & -1 & -1 \\ 2 & -1 & 0 \\ 0 & 0 & 1 \end{bmatrix}$$

Letting $x = T_o z$ yields

$$\dot{z} = \begin{bmatrix} \dot{z}_o \\ \dot{z}_{\bar{o}} \end{bmatrix} = \left[\begin{array}{cc:c} 0 & 1 & 0 \\ -2 & 3 & 0 \\ \hdashline -5 & 3 & 2 \end{array}\right] \begin{bmatrix} z_o \\ z_{\bar{o}} \end{bmatrix} + \begin{bmatrix} 1 \\ 2 \\ 1 \end{bmatrix} u$$

$$y = [1 \quad 0 \;\vdots\; 0] \begin{bmatrix} z_o \\ z_{\bar{o}} \end{bmatrix}$$

3.7.3 Canonical Decomposition Theory

Consider the linear time-invariant system

$$\dot{x}(t) = Ax(t) + Bu(t)$$
$$y(t) = Cx(t)$$

By nonsingular transformation, (A, B, C) can be transformed into the following canonical form

$$\left.\begin{aligned} \begin{bmatrix} \dot{w}_{co} \\ \dot{w}_{c\bar{o}} \\ \dot{w}_{\bar{c}o} \\ \dot{w}_{\bar{c}\bar{o}} \end{bmatrix} &= \begin{bmatrix} \hat{A}_{co} & 0 & \hat{A}_{13} & 0 \\ \hat{A}_{21} & \hat{A}_{c\bar{o}} & \hat{A}_{23} & \hat{A}_{24} \\ 0 & 0 & \hat{A}_{\bar{c}o} & 0 \\ 0 & 0 & \hat{A}_{43} & \hat{A}_{\bar{c}\bar{o}} \end{bmatrix} \begin{bmatrix} w_{co} \\ w_{c\bar{o}} \\ w_{\bar{c}o} \\ w_{\bar{c}\bar{o}} \end{bmatrix} + \begin{bmatrix} \hat{B}_{co} \\ \hat{B}_{c\bar{o}} \\ 0 \\ 0 \end{bmatrix} u \\ y &= [\hat{C}_{co} \quad 0 \quad \hat{C}_{\bar{c}o} \quad 0] \begin{bmatrix} w_{co} \\ w_{c\bar{o}} \\ w_{\bar{c}o} \\ w_{\bar{c}\bar{o}} \end{bmatrix} \end{aligned}\right\} \tag{3.39}$$

where w_{co} is controllable and observable, $w_{c\bar{o}}$ controllable but not observable, $w_{\bar{c}o}$ is uncontrollable but observable, and $w_{\bar{c}\bar{o}}$ is neither controllable nor observable.

Proof If the system (A,B,C) is uncontrollable, it can be transformed into the form of Eq. (3.33). Consider now the controllable part and uncontrollable part of (A,B,C), respectively. If the controllable part and/or the uncontrollable are unobservable, then they can be transformed into the form of Eq.(3.36), respectively. Combining these above transformation, we obtain equation (3.39). Furthermore, the transfer function of the system (A,B,C) is

$$G(s) = \hat{C}_{co}(sI - \hat{A}_{co})^{-1}\hat{B}_{co} \tag{3.40}$$

Problems

P3.1 Check the state controllability and observability of the following systems.

(a) $A = \begin{bmatrix} 1 & 0 \\ -1 & 2 \end{bmatrix}$, $b = \begin{bmatrix} 1 \\ 0 \end{bmatrix}$, $c = [0 \ 1]$

(b) $A = \begin{bmatrix} -2 & 2 & -1 \\ 0 & -2 & 0 \\ 1 & -4 & 0 \end{bmatrix}$, $b = \begin{bmatrix} 0 \\ 0 \\ 1 \end{bmatrix}$, $c = [1 \ -1 \ 1]$

(c) $A = \begin{bmatrix} -3 & 1 & 0 \\ 0 & -3 & 0 \\ 0 & 0 & -1 \end{bmatrix}$, $B = \begin{bmatrix} 1 & -1 \\ 0 & 0 \\ 3 & 0 \end{bmatrix}$, $C = \begin{bmatrix} 1 & 0 & 0 \\ 0 & 0 & 1 \end{bmatrix}$

(d) $A = \begin{bmatrix} -5 & 0 & 0 \\ 0 & -5 & 0 \\ 0 & 0 & 1 \end{bmatrix}$, $b = \begin{bmatrix} 1 \\ 3 \\ 2 \end{bmatrix}$, $C = \begin{bmatrix} 1 & 2 & 3 \\ 2 & 4 & 0 \end{bmatrix}$

(e) $A = \begin{bmatrix} -2 & 0 & 0 & 0 \\ 0 & -5 & 1 & 0 \\ 0 & 0 & -5 & 1 \\ 0 & 0 & 0 & -5 \end{bmatrix}$, $b = \begin{bmatrix} 2 \\ 0 \\ 0 \\ 1 \end{bmatrix}$, $c = [1 \ 1 \ 0 \ 0]$

(f) $A = \begin{bmatrix} -2 & 0 & 0 & 0 \\ 0 & -5 & 1 & 0 \\ 0 & 0 & -5 & 1 \\ 0 & 0 & 0 & -5 \end{bmatrix}$, $b = \begin{bmatrix} 1 \\ 2 \\ 0 \\ 0 \end{bmatrix}$, $c = [1 \ 0 \ 1 \ 0]$

P3.2 Consider the system

$$\dot{x} = \begin{bmatrix} a & 1 \\ 0 & b \end{bmatrix}x + \begin{bmatrix} 1 \\ 1 \end{bmatrix}u, \quad y = [1 \ -1]x$$

Find the region in the a-versus-b plane such that the system is state controllable and observable.

P3.3 Determine whether the following systems are output controllable.

(a) $A = \begin{bmatrix} 0 & 1 & 0 \\ 0 & 0 & 1 \\ -6 & -11 & -6 \end{bmatrix}$, $b = \begin{bmatrix} 0 \\ 0 \\ 1 \end{bmatrix}$, $c = [1 \ 0 \ 0]$

(b) $\boldsymbol{A} = \begin{bmatrix} -a & 0 & 0 & 0 \\ 0 & -b & 0 & 0 \\ 0 & 0 & -c & 0 \\ 0 & 0 & 0 & -d \end{bmatrix}$, $\boldsymbol{b} = \begin{bmatrix} 0 \\ 0 \\ 1 \\ 1 \end{bmatrix}$, $\boldsymbol{c} = [1 \ \ 0 \ \ 0 \ \ 0]$

P3.4 Consider the state-space representation

$$\dot{\boldsymbol{x}} = \begin{bmatrix} -1 & 0 \\ 1 & -2 \end{bmatrix} \boldsymbol{x} + \begin{bmatrix} 1 \\ -1 \end{bmatrix} u,$$
$$y = [-1 \ \ 1] \boldsymbol{x}$$

(a) Transform the given representation into the controllable canonical form by a suitable transformation $\boldsymbol{x} = \boldsymbol{P}_c \boldsymbol{z}$.

(b) Transform the given representation into the observable canonical form by a suitable transformation $\boldsymbol{x} = \boldsymbol{P}_o \boldsymbol{z}$.

P3.5 A control system has the transfer function of

$$G(s) = \frac{s + a}{s^3 + 7s^2 + 14s + 8}$$

Find the values of a such that the system is uncontrollable and/or unobservable.

P3.6 A control system has the transfer function of

$$G(s) = \frac{s + 1}{s^2 + 3s + 2}$$

find the state-space representations so such that the system is (a) controllable but unobservable; (b) observable but uncontrollable; (c) neither controllable nor observable.

P3.7 Consider the system

$$\dot{\boldsymbol{x}} = \begin{bmatrix} 1 & 0 & 0 & 0 \\ 0 & 2 & 0 & 0 \\ -6 & -2 & 3 & 0 \\ 3 & -2 & 0 & 4 \end{bmatrix} \boldsymbol{x} + \begin{bmatrix} 1 \\ 0 \\ 3 \\ 1 \end{bmatrix} u,$$
$$y = [-4 \ \ -3 \ \ 1 \ \ 1] \boldsymbol{x}$$

(a) Find the controllable subsystem and the uncontrollable subsystem.

(b) Find the observable subsystem and the unobservable subsystem.

(c) determine the transfer function of the system.

P3.8 Determine the controllability of the following systems. Transform the controllable systems into the controllable canonical form, and transform the uncontrollable systems into the standard uncontrollable form.

(a) $\boldsymbol{A} = \begin{bmatrix} -1 & 0 \\ 0 & -2 \end{bmatrix}$, $\boldsymbol{c} = [-1 \ \ 1]$

(b) $\boldsymbol{A} = \begin{bmatrix} -2 & 2 \\ 0 & 24 \end{bmatrix}$, $\boldsymbol{c} = [1 \ \ 1]$

(c) $A = \begin{bmatrix} 1 & 2 & 0 \\ 3 & -1 & 1 \\ 0 & 2 & 0 \end{bmatrix}$, $b = \begin{bmatrix} 1 \\ 0 \\ 1 \end{bmatrix}$

(d) $A = \begin{bmatrix} 1 & 0 & 0 \\ 2 & 2 & 3 \\ -1 & 0 & 1 \end{bmatrix}$, $b = \begin{bmatrix} 1 \\ 2 \\ -2 \end{bmatrix}$

P3.9 Determine the observability of the following systems. Transform the observable systems into the observable canonical form, and transform the unobservable systems into the standard unobservable form.

(a) $A = \begin{bmatrix} 1 & 0 \\ -2 & 4 \end{bmatrix}$, $b = \begin{bmatrix} 1 \\ 1 \end{bmatrix}$

(b) $A = \begin{bmatrix} -1 & 0 \\ 1 & -2 \end{bmatrix}$, $b = \begin{bmatrix} 1 \\ 1 \end{bmatrix}$

(c) $A = \begin{bmatrix} 2 & 0 & 0 \\ 0 & 4 & 1 \\ 0 & 0 & 4 \end{bmatrix}$, $c = [0 \;\; 0 \;\; 1]$

Chapter 4 State-Space Synthesis of Linear Systems

4.1 State Feedback and Output Feedback

So far, in our study of state-apace method we are involved only open-loop systems as shown in Fig. 4.1, where it is assumed that the matrix $\boldsymbol{D} = \boldsymbol{0}$.

Consider a n-dimensional linear time-invariant continuous system

$$\left.\begin{aligned}\dot{\boldsymbol{x}}(t) &= \boldsymbol{A}\boldsymbol{x}(t) + \boldsymbol{B}\boldsymbol{u}(t)\\ \boldsymbol{y}(t) &= \boldsymbol{C}\boldsymbol{x}(t)\end{aligned}\right\} \qquad (4.1)$$

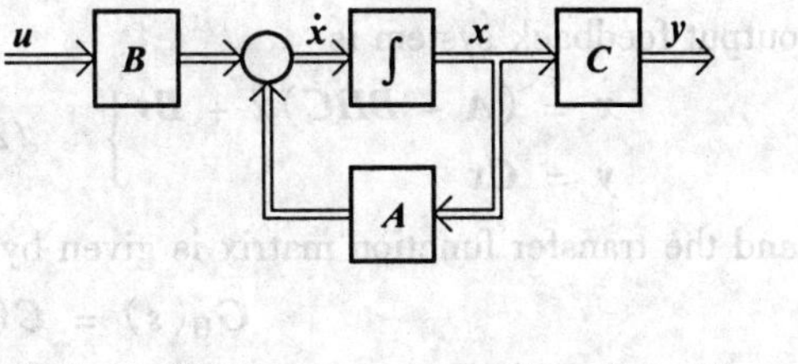

Figure 4.1 State space representation of open-loop system

where $\boldsymbol{x}, \boldsymbol{u}, \boldsymbol{y}$ are $n \times 1, p \times 1, q \times 1$ vectors, respectively. Since the behavior of a system is now described by the state variables, it is reasonable to choose the control $\boldsymbol{u}$ based on the state $\boldsymbol{x}$ and reference input $\boldsymbol{v}$, i.e., a good control should be determined by an equation of the form $\boldsymbol{u}(t) = f(\boldsymbol{v}(t), \boldsymbol{x}(t))$. This relation is called the control law.

For designing control systems, however, feedback is an essential manner in both classical control theory and modern control theory. When a system is described by the state variables, there are two kinds of feedback for the feedback control systems, the state feedback and the output feedback.

4.1.1 State Feedback

In state feedback, every state is multiplied by a gain and fed back to the input terminal as shown in Fig. 4.2, where $\boldsymbol{F}$ is the feedback gain matrix.

At first, let's investigate the effect of the introduced state feedback on the state-space representation of a system. In Fig. 4.2, the control law is determined by the linear equation

$$\boldsymbol{u} = \boldsymbol{v} - \boldsymbol{F}\boldsymbol{x} \qquad (4.2)$$

Substituting Eq. (4.2) into Eq. (4.1) yields the state-space representation of the state feedback system as

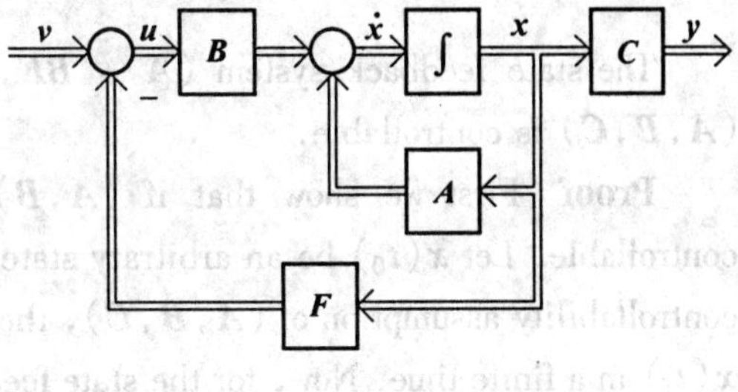

Figure 4.2 State feedback system

$$\left.\begin{aligned}\dot{\boldsymbol{x}}(t) &= (\boldsymbol{A} - \boldsymbol{B}\boldsymbol{F})\boldsymbol{x}(t) + \boldsymbol{B}\boldsymbol{v}(t)\\ \boldsymbol{y} &= \boldsymbol{C}\boldsymbol{x}(t)\end{aligned}\right\} \qquad (4.3)$$

and the closed-loop transfer function matrix is given by

$$G_F(s) = C(sI - A + BF)^{-1}B \tag{4.4}$$

4.1.2 Output Feedback

In engineering practice, output feedback sometimes is adopted, especially when not all state variables are measurable. A typical output feedback is shown in Fig. 4.3, where H is the feedback gain matrix.

It can be seen, now, the linear feedback law is

$$u = v - Hy = v - HCx \tag{4.5}$$

Then, the state-space representation of the output feedback system is

$$\left.\begin{aligned} \dot{x} &= (A - BHC)x + Bv \\ y &= Cx \end{aligned}\right\} \tag{4.6}$$

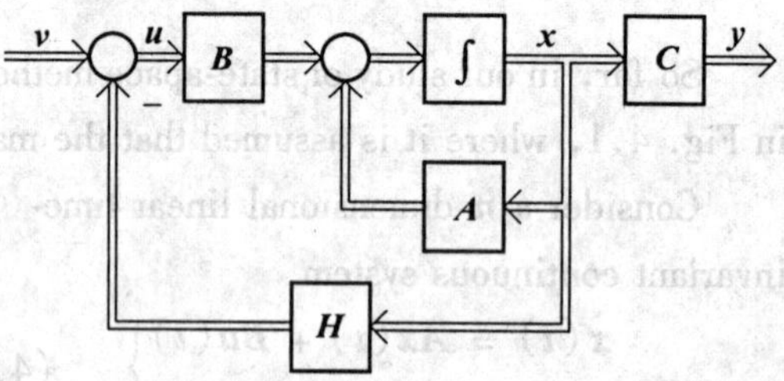

Fig. 4.3 Output feedback

and the transfer function matrix is given by

$$G_H(s) = C(sI - A + BHC)^{-1}B \tag{4.7}$$

It is useful for us to understand the basic characters of state feedback and output feedback and distinction between them.

(a) Both the state feedback and output feedback don't introduce any new state variable, i.e., they don't change the dimension of their original open-loop systems.

(b) Since the number of state variables is generally larger than that of output variables, what can be achieved by output feedback can always be achieved by state feedback, but the converse is not true.

(c) In practice, it is usually difficult to sense all the states and feed them back. In reality only certain states or combinations of them are measurable as outputs. Consequently, any practical compensator must rely on systems outputs, inputs, and a few state variables for compensation.

4.1.3 Controllability of State Feedback System

The state feedback system $(A - BF, B, C)$ is controllable if and only if the system (A, B, C) is controllable.

Proof First we show that if (A, B) is controllable then $(A - BF, B)$ must be controllable. Let $x(t_0)$ be an arbitrary state of the system (A, B, C) and $x(t_f) = \mathbf{0}$. By the controllability assumption of (A, B, C), there exists an input $u(t)$ that will transfer $x(t_0)$ to $x(t_f)$ in a finite time. Now, for the state feedback system $(A - BF, B, C)$, if we choose $v = u + Fx$, then the input $v(t)$ will transfer $x(t_0)$ to $x(t_f)$ in a finite time. Therefore, the state feedback system $(A - BF, B, C)$ must be controllable.

On the other hand, from Fig. 4.2 it can be seen that the control controlling state is produced from the input $\boldsymbol{v}(t)$. Hence, if $\boldsymbol{u}(t)$ cannot control $\boldsymbol{x}(t)$, neither can $\boldsymbol{v}(t)$; i.e., if $(\boldsymbol{A},\boldsymbol{B},\boldsymbol{C})$ is not controllable, neither is $(\boldsymbol{A}-\boldsymbol{BF},\boldsymbol{B},\boldsymbol{C})$.

4.2 Pole Assignment Using State Feedback

Since both state feedback and output feedback change the system matrices of the closed-loop system, they can be used to change the pole positions of a system. Especially, the state feedback can be used to achieve the desired pole positions.

4.2.1 Condition of Pole Assignment

If a single-input system $(\boldsymbol{A},\boldsymbol{B},\boldsymbol{C})$ is controllable then the eigenvalues of $(\boldsymbol{A}-\boldsymbol{BF},\boldsymbol{B},\boldsymbol{C})$, i.e. the poles of the state feedback system, can be assigned arbitrarily.

Proof If a single-input system $(\boldsymbol{A},\boldsymbol{B},\boldsymbol{C})$ is controllable, then it can be transformed into the controllable canonical $(\boldsymbol{A}_c,\boldsymbol{B}_c)$ by a suitable transformation $\boldsymbol{x}=\boldsymbol{P}_c\boldsymbol{z}$, i.e.,

$$\boldsymbol{A}_c=\boldsymbol{P}_c^{-1}\boldsymbol{AP}_c=\begin{bmatrix}0 & 1 & \cdots & 0\\ \vdots & \vdots & & \vdots\\ 0 & 0 & \cdots & 1\\ -\alpha_n & -\alpha_{n-1} & \cdots & -\alpha_1\end{bmatrix},\quad \boldsymbol{B}_c=\boldsymbol{P}_c^{-1}\boldsymbol{B}=\begin{bmatrix}0\\ \vdots\\ 0\\ 1\end{bmatrix}$$

$$\boldsymbol{C}_c=\boldsymbol{CP}_c=[\beta_n\quad \beta_{n-1}\quad \cdots\quad \beta_1]$$

where α_i and β_i, $i=1,2,\cdots,n$, are constant coefficients. By inspection, the transfer function of the open-loop system is

$$\frac{Y(s)}{U(s)}=\boldsymbol{C}(s\boldsymbol{I}-\boldsymbol{A})^{-1}\boldsymbol{B}=\boldsymbol{C}_c(s\boldsymbol{I}-\boldsymbol{A}_c)^{-1}\boldsymbol{B}_c=\frac{\beta_1 s^{n-1}+\cdots+\beta_{n-1}s+\beta_n}{s^n+\alpha_1 s^{n-1}+\cdots+\alpha_{n-1}s+\alpha_n}\tag{4.8}$$

Introducing state feedback

$$u=v-\boldsymbol{Fx}=v-\boldsymbol{FP}_c\boldsymbol{z}=v-\boldsymbol{F}_z\boldsymbol{z}\tag{4.9}$$

where

$$\boldsymbol{F}_z=\boldsymbol{FP}_c=[f_{z1}\quad f_{z2}\quad \cdots\quad f_{zn}]\tag{4.10}$$

The system matrix of the state feedback system is

$$\boldsymbol{A}_c-\boldsymbol{B}_c\boldsymbol{F}_z=\begin{bmatrix}0 & 1 & \cdots & 0\\ \vdots & \vdots & & \vdots\\ 0 & 0 & \cdots & 1\\ -\alpha_n-f_{z1} & -\alpha_{n-1}-f_{z2} & \cdots & -\alpha_1-f_{zn}\end{bmatrix}\tag{4.11}$$

The transfer function of the closed-loop system, for this canonical form, is

$$\frac{Y(s)}{V(s)}=\frac{\beta_1 s^{n-1}+\cdots+\beta_{n-1}s+\beta_n}{s^n+(\alpha_1+f_{zn})s^{n-1}+\cdots+(\alpha_{n-1}+f_{z2})s+(\alpha_n+f_{z1})}\tag{4.12}$$

Since the characteristic polynomial is

$$\Delta(s)=s^n+(\alpha_1+f_{zn})s^{n-1}+\cdots+(\alpha_{n-1}+f_{z2})s+(\alpha_n+f_{z1})\tag{4.13}$$

we can conclude that the eigenvalues of the matrix $(\boldsymbol{A}_c-\boldsymbol{B}_c\boldsymbol{F}_z)$, i.e. the poles of the closed-loop system, can be assigned arbitrarily by choosing appropriate state feedback matrix F_z. Then, the state feedback matrix F can be calculated from F_z.

4.2.2 Algorithm of Pole Assignment

1. Direct Comparing Characteristic Polynomials

Assume that the system (A, B, C) is controllable, then the characteristic polynomial of the state feedback system $(A - BF, B, C)$ is

$$\Delta(s) = |sI - A + BF| \tag{4.14}$$

Given the desired poles of the closed-loop system $\hat{s}_1, \hat{s}_2, \cdots, \hat{s}_n$, the required characteristic polynomial should be as

$$\Delta_d(s) = (s - \hat{s}_1)(s - \hat{s}_2)\cdots(s - \hat{s}_n) = s^n + \hat{\alpha}_1 s^{n-1} + \cdots + \hat{\alpha}_{n-1}s + \hat{\alpha}_n \tag{4.15}$$

The elements of state feedback matrix F can be obtained by comparing the coefficients of Eq. (4.14) and Eq. (4.15).

Example 4.1 Consider the system given by

$$\dot{x} = \begin{bmatrix} 1 & 0 \\ -1 & 2 \end{bmatrix} x + \begin{bmatrix} -1 \\ 1 \end{bmatrix} u$$

Find the state feedback matrix F so that the closed-loop poles are located at $\hat{s}_{1,2} = -1 \pm j2$.

Solution At first, examine the controllability of the open-loop system. Since

$$\text{rank } Q_c = \text{rank}\begin{bmatrix} -1 & -1 \\ 1 & 3 \end{bmatrix} = 2$$

thus, the system is controllable and the poles of the state feedback system can be assigned arbitrarily.

Denoting $F = [f_1 \quad f_2]$ yields

$$BF = \begin{bmatrix} -1 \\ 1 \end{bmatrix}[f_1 \quad f_2] = \begin{bmatrix} -f_1 & -f_2 \\ f_1 & f_2 \end{bmatrix}$$

$$sI - A + BF = \begin{bmatrix} s & 0 \\ 0 & s \end{bmatrix} - \begin{bmatrix} 1 & 0 \\ -1 & 2 \end{bmatrix} + \begin{bmatrix} -f_1 & -f_2 \\ f_1 & f_2 \end{bmatrix} = \begin{bmatrix} s - 1 - f_1 & -f_2 \\ 1 + f_1 & s - 2 + f_2 \end{bmatrix}$$

Taking the determinant of the matrix yields the characteristic polynomial

$$|sI - (A - BF)| = s^2 + (-f_1 + f_2 - 3)s + (2f_1 + 2)$$

The desired characteristic polynomial is

$$\Delta_d(s) = (s - \hat{s}_1)(s - \hat{s}_2) = s^2 + 2s + 5$$

Equating the coefficients of theses two equations get

$$-f_1 + f_2 - 3 = 2$$

$$2 + 2f_1 = 5$$

This results in $f_1 = 1.5$ and $f_2 = 6.5$. Hence, the feedback gain matrix is

$$F = [1.5 \quad 6.5]$$

This method, though quite simple for lower-order systems, may become cumbersome for systems of more than third-order. For such systems, alternative method is better suited.

2. Using Canonical Form

It can be seen that subsection 4.2.1 gives another method, via the controllable canonical form, for determine the state feedback matrix to place the closed-loop poles at the desired position. Since the matrix $\boldsymbol{F}_z$ is chosen with respect to the state z, we have to use $\boldsymbol{F} = \boldsymbol{F}_z \boldsymbol{P}_c^{-1}$, given by Eq. (4.10).

Given a controllable system $(\boldsymbol{A}, \boldsymbol{B}, \boldsymbol{C})$ and a set of desired poles $\hat{s}_1, \hat{s}_2, \cdots, \hat{s}_n$, the procedure of determining state feedback matrix $\boldsymbol{F}$ is summarized as follows.

(1) Find the characteristic polynomial of matrix $\boldsymbol{A}$

$$\det(s\boldsymbol{I} - \boldsymbol{A}) = s^n + \alpha_1 s^{n-1} + \cdots + \alpha_{n-1} s + \alpha_n$$

(2) Calculate the required characteristic polynomial

$$\Delta_d(s) = (s - \hat{s}_1)(s - \hat{s}_2)\cdots(s - \hat{s}_n) = s^n + \hat{\alpha}_1 s^{n-1} + \cdots + \hat{\alpha}_{n-1} s + \hat{\alpha}_n$$

(3) Calculate the state feedback matrix with respect to state z

$$\boldsymbol{F}_z = [f_{z1} \quad f_{z2} \quad \cdots \quad f_{zn}] = [\hat{\alpha}_n - \alpha_n \quad \hat{\alpha}_{n-1} - \alpha_{n-1} \quad \cdots \quad \hat{\alpha}_1 - \alpha_1]$$

(4) Calculate the transformation matrix $\boldsymbol{P}_c$ transforming $(\boldsymbol{A}, \boldsymbol{B})$ into the controllable canonical form, using Eq. (3.11), and its inverse $\boldsymbol{P}_c^{-1}$.

(5) Calculate the required state feedback matrix

$$\boldsymbol{F} = \boldsymbol{F}_z \boldsymbol{P}_c^{-1}$$

Example 4.2 Again consider the system given in Example 4.1

$$\dot{\boldsymbol{x}} = \begin{bmatrix} 1 & 0 \\ -1 & 2 \end{bmatrix} \boldsymbol{x} + \begin{bmatrix} -1 \\ 1 \end{bmatrix} u$$

Find the state feedback matrix $\boldsymbol{F}$ such that the closed-loop poles are located at $\hat{s}_{1,2} = -1 \pm \mathrm{j}2$, via the controllability canonical form of the state-space representation.

Solution From Example 4.1, the given system is controllable.

(1) The characterized polynomial of $\boldsymbol{A}$ matrix is

$$\det(s\boldsymbol{I} - \boldsymbol{A}) = (s - 1)(s - 2) = s^2 - 3s + 2$$

(2) The required characteristic polynomial is

$$\Delta_d(s) = (s - \hat{s}_1)(s - \hat{s}_2) = s^2 + 2s + 5$$

(3) The state feedback matrix with respect to state z is

$$\boldsymbol{F}_z = [f_{z1} \quad f_{z2}] = [\alpha_2 - \alpha_2 \quad \hat{\alpha}_1 - \alpha_1] = [3 \quad 5]$$

(4) The transformation matrix $\boldsymbol{P}_c$ and its inverse are

$$\boldsymbol{P}_c = \begin{bmatrix} -1 & -1 \\ 1 & 3 \end{bmatrix} \begin{bmatrix} -3 & 1 \\ 1 & 0 \end{bmatrix} = \begin{bmatrix} 2 & -1 \\ 0 & 1 \end{bmatrix}, \quad \boldsymbol{P}_c^{-1} = \begin{bmatrix} 0.5 & 0.5 \\ 0 & 1 \end{bmatrix}$$

(5) The required state feedback matrix is

$$\boldsymbol{F} = \boldsymbol{F}_z \boldsymbol{P}_c^{-1} = [3 \quad 5] \begin{bmatrix} 0.5 & 0.5 \\ 0 & 1 \end{bmatrix} = [1.5 \quad 6.5]$$

State feedback is not usually practical for two reasons. The first is that state feedback leads to PD or PID compensators, which have wider bandwidth. The second reason is that it is simply not possible or practical to sense all the states and feed them back. In reality only certain states or combinations of them are measurable as outputs. Consequently any practical compensator must rely only on system outputs, inputs, and a few state variables.

4.3 Stabilization

If a system $(\boldsymbol{A},\boldsymbol{B},\boldsymbol{C})$ is controllable, then all the eigenvalues of the system matrix i.e., all the poles of the state feedback system, can be arbitrarily assigned by introducing state feedback. If a system $(\boldsymbol{A},\boldsymbol{B},\boldsymbol{C})$ is not complete state controllable, then the state equation can be decomposed, according to the controllability, into the form

$$\begin{bmatrix} \dot{z}_c \\ \dot{z}_{\bar{c}} \end{bmatrix} = \begin{bmatrix} \overline{\boldsymbol{A}}_c & \overline{\boldsymbol{A}}_{12} \\ 0 & \overline{\boldsymbol{A}}_{\bar{c}} \end{bmatrix} \begin{bmatrix} z_c \\ z_{\bar{c}} \end{bmatrix} + \begin{bmatrix} \overline{\boldsymbol{B}}_c \\ 0 \end{bmatrix} \boldsymbol{u} \tag{4.16}$$

and the reduced subsystem

$$\dot{z}_c = \overline{\boldsymbol{A}}_c z_c + \overline{\boldsymbol{B}}_c \boldsymbol{u} \tag{4.17}$$

is controllable. Because of the form of the system matrix, its eigenvalues are the union of the eigenvalues of $\overline{\boldsymbol{A}}_c$ and $\overline{\boldsymbol{A}}_{\bar{c}}$. In view of the form of the input matrix, it is easy to see that all the eigenvalues of $\overline{\boldsymbol{A}}_{\bar{c}}$ cannot be controlled. On the other hand, all the eigenvalues of $\overline{\boldsymbol{A}}_c$ can be arbitrarily assigned by introducing state feedback.

In the design of a system, sometimes it is required only to change unstable eigenvalues (the eigenvalues with nonnegative real parts) to stable eigenvalues (the eigenvalues with negative real parts). This is called stabilization. From the above discussion it can be seen that if the matrix $\overline{\boldsymbol{A}}_{\bar{c}}$ has unstable eigenvalues, then the system cannot be stabilized.

4.4 State Observer

In the previous discussion the state feedback is introduced under the assumption that all the state variables are measurable and available as outputs. This assumption does not always hold in practice, however, because of the technical or economic reasons. Thus, in order to apply state feedback to a system, a reasonable substitute for the states often has to be found. We will now discuss the problem of actually obtaining the states of the system from knowledge only of the system inputs and outputs, and developing a realistic state estimator, which is usually known as asymptotic observer.

In the discussion of state observer, the subscript e with a variable is used to denote an estimate of the variable. For example, state variable $\boldsymbol{x}_e$ is an estimate of $\boldsymbol{x}$.

4.4.1 State Observer

1. Open-loop Observer

Consider the linear time-invariant system $(\boldsymbol{A},\boldsymbol{B},\boldsymbol{C})$. It is assumed that the state variables are not available. Since the matrices $\boldsymbol{A}$, $\boldsymbol{B}$ and $\boldsymbol{C}$ are known, the problem is to estimate or generate the state variables from the available inputs and outputs with the knowledge of $\boldsymbol{A}$, $\boldsymbol{B}$ and $\boldsymbol{C}$. A simply way is to duplicate the original system, using integrated circuit or digital computer, as shown in Fig. 4.4. This duplicated system is called an open-loop observer.

If the original system and the observer have the same initial state and the same input, the states $\boldsymbol{x}_e(t)$ of the observer will be equal to the actual state $\boldsymbol{x}(t)$ for all time. Note that the output is no longer required. Unfortunately, however, this fact is really more of disadvantages than help. The point is that this strategy is susceptible to disturbances that may arise during system operation; and allow no means of compensation for any error, possibly small but still almost inevitable, in the predetermined strategy.

Assume that the initial state of the observer is slightly in error,

$$\boldsymbol{x}(t_0) - \boldsymbol{x}_e(t_0) = \varepsilon$$

Then, since the original system and the observer have the same input, we have

$$\dot{\boldsymbol{x}}(t) - \dot{\boldsymbol{x}}_e(t) = \boldsymbol{A}[\boldsymbol{x}(t) - \boldsymbol{x}_e(t)]$$

and

$$\boldsymbol{x}(t) - \boldsymbol{x}_e(t) = e^{\boldsymbol{A}t}[\boldsymbol{x}(t_0) - \boldsymbol{x}_e(t_0)] = e^{\boldsymbol{A}t} \cdot \varepsilon$$

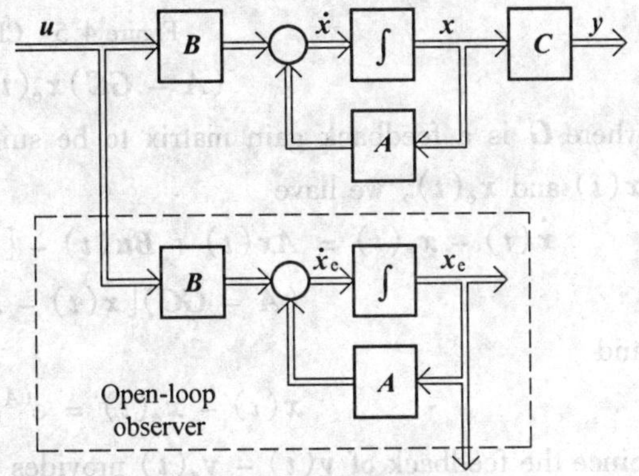

Figure 4.4 Open-loop observer

It can be seen that if the matrix $\boldsymbol{A}$ has eigenvalues with positive real parts, i.e., the system is unstable, the error between the actual state $\boldsymbol{x}(t)$ and the estimated state $\boldsymbol{x}_e(t)$ will increase with time no matter how small the initial error is. Even the system is stable but the matrix $\boldsymbol{A}$ has some eigenvalues with very small real parts, the effect of the initial error will take a long time to die out. Therefore the open-loop observer is, in general, not very satisfactory.

2. Closed-loop Observer

The classical way of overcoming the potential difficulties of open-loop observer is to use feedback, i.e., to try to zero the error by driving the system with a term proportional to the error in the estimate. But there is a problem of how to obtain the error signal, because $\boldsymbol{x}(t)$ is not available for us. It can be noted that the system output was not used in the open-loop observer; and that the output, which is closed related to the state, is always available. Therefore, an error signal can be generated as

$$\boldsymbol{y}(t) - \boldsymbol{y}_e(t) = \boldsymbol{C}[\boldsymbol{x}(t) - \boldsymbol{x}_e(t)]$$

and it can be used to control the previous open-loop observer. This consideration leads us to construct an asymptotic state observer as shown in Fig. 4.5.

From Fig. 4.5, the state equation of the closed-loop observer is

$$\dot{\boldsymbol{x}}_e(t) = \boldsymbol{A}\boldsymbol{x}_e(t) + \boldsymbol{B}\boldsymbol{u}(t) + \boldsymbol{G}[\boldsymbol{y}(t) - \boldsymbol{C}\boldsymbol{x}_e(t)] =$$

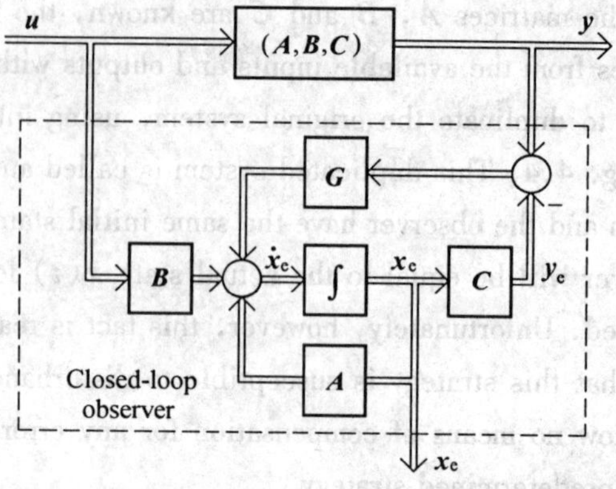

Figure 4.5 Closed-loop observer

$$(A - GC)x_e(t) + Bu(t) + Gy(t) \tag{4.18}$$

where G is a feedback gain matrix to be suitably chosen. Again consider the error between $x(t)$ and $x_e(t)$, we have

$$\dot{x}(t) - \dot{x}_e(t) = Ax(t) + Bu(t) - [(A - GC)x_e(t) + Bu(t) + Gy(t)] = (A - GC)[x(t) - x_e(t)] \tag{4.19}$$

and

$$x(t) - x_e(t) = e^{(A-GC)t}[x(t_0) - x_e(t_0)] \tag{4.20}$$

Since the feedback of $y(t) - y_e(t)$ provides some control over the behavior of the error $x(t) - x_e(t)$, if all the eigenvalues of $(A - GC)$ have negative real parts, then no matter what the initial state $x_e(t_0)$ of the observer is, the error $x(t) - x_e(t)$ will approach zero asymptotically, i.e., the estimated $x_e(t)$ will approach the actual state $x(t)$ asymptotically. This is the reason why this observer is called asymptotic observer. In short, if the eigenvalues of $(A - GC)$ can be chosen properly, an asymptotic state observer is much more desirable than an open-loop state observer.

4.4.2 Design of State Observer

1. Condition of Assigning Observer's Poles

It can be seen, from Eq. (4.20), that if all the eigenvalues of $(A - GC)$ have negative real parts, then

$$\lim_{t\to\infty}[x(t) - x_e(t)] = 0$$

and the rate of decay is determined by the eigenvalues of $(A - GC)$. The question is that can we always find such a suitable matrix G? The answer is that if (A, C) is observable then the eigenvalues of $(A - GC)$ can be assigned arbitrarily.

Proof According to the principle of duality, If (A, C) is observable then (A^T, C^T) must be controllable, and there exists a gain matrix F so that the eigenvalues of $(A^T - C^TF)$ can be assigned arbitrarily. Since the transpose does not change the eigenvalues of a matrix,

$(\boldsymbol{A}-\boldsymbol{F}^{\mathrm{T}}\boldsymbol{C})$ has the same eigenvalues. Hence, letting $\boldsymbol{G}=\boldsymbol{F}^{\mathrm{T}}$, if $(\boldsymbol{A},\boldsymbol{C})$ is observable then the eigenvalues of $(\boldsymbol{A}-\boldsymbol{GC})$ can be assigned arbitrarily.

It should be pointed, however, that it should be taken into consideration whether or not the eigenvalues with large negative real parts will make the system more susceptible to disturbances. Hence the best eigenvalues of $(\boldsymbol{A}-\boldsymbol{GC})$, i.e., the poles of an observer should be the compromise between the response rapidness and anti-noise of the observer.

2. Algorithm for Observer Gain Matrix

This problem of assigning the poles of an observer is essentially the same as the one we discussed in determining the state feedback matrix $\boldsymbol{F}$ due to the duality between the controllability and observability.

(1) Direct comparing characteristic polynomials

If a system $(\boldsymbol{A},\boldsymbol{B},\boldsymbol{C})$ is observable, then the state equation of the observer is, from Eq. (4.18) is given by

$$\dot{\boldsymbol{x}}_{\mathrm{e}}(t)=(\boldsymbol{A}-\boldsymbol{GC})\boldsymbol{x}_{\mathrm{e}}(t)+\boldsymbol{Bu}(t)+\boldsymbol{Gy}(t)$$

One way of obtaining the feedback gain matrix $\boldsymbol{G}$ is to compare the characteristic polynomial of the observer

$$\Delta(s)=|s\boldsymbol{I}-(\boldsymbol{A}-\boldsymbol{GC})| \tag{4.21}$$

with the desired characteristic polynomial

$$\Delta_{\mathrm{d}}(s)=(s-\hat{s}_1)(s-\hat{s}_2)\cdots(s-\hat{s}_n)=s^n+\hat{\alpha}_1 s^{n-1}+\cdots+\hat{\alpha}_{n-1}s+\hat{\alpha}_n \tag{4.22}$$

where $\hat{s}_1,\hat{s}_2,\cdots,\hat{s}_n$ are the desired poles of the closed-loop system.

Example 4.3 Consider the system described by

$$\dot{\boldsymbol{x}}=\begin{bmatrix}1&0&0\\3&-1&1\\0&2&0\end{bmatrix}\boldsymbol{x}+\begin{bmatrix}2\\1\\1\end{bmatrix}u,$$

$$y=\begin{bmatrix}0&0&1\end{bmatrix}\boldsymbol{x}$$

Design an asymptotic observer such that its poles are located at $\hat{s}_1=-3$, $\hat{s}_2=-4$, and $\hat{s}_3=-5$.

Solution Checking the observable matrix of the given system, we get

$$\operatorname{rank}\boldsymbol{Q}_{\mathrm{o}}=\operatorname{rank}\begin{bmatrix}0&0&1\\0&2&0\\6&-2&2\end{bmatrix}=3$$

Thus, the system is observable and the poles of the observer can be assigned arbitrarily.

Denoting $\boldsymbol{G}=[g_1\quad g_2\quad g_3]^{\mathrm{T}}$, we get

$$|s\boldsymbol{I}-\boldsymbol{A}+\boldsymbol{GC}|=\left|\begin{bmatrix}s-1&0&0\\-3&s+1&-1\\0&-2&s\end{bmatrix}+\begin{bmatrix}0&0&g_1\\0&0&g_2\\0&0&g_3\end{bmatrix}\right|=\begin{vmatrix}s-1&0&g_1\\-3&s+1&-1+g_2\\0&-2&s+g_3\end{vmatrix}=$$

$$s^3+g_3s^2+(2g_2-3)s+(6g_1-2g_2-g_3+2)$$

The required characteristic polynomial is

$$\Delta_d(s) = (s+3)(s+4)(s+5) = s^3 + 12s^2 + 47s + 2$$

Equating the coefficients of theses two equations, we have

$$G = [g_1 \quad g_2 \quad g_3]^T = [20 \quad 25 \quad 12]^T$$

The state equation of the observer is

$$\dot{x}_e(t) = (A - GC)x_e(t) + Bu(t) + Gy(t) = \begin{bmatrix} 1 & 0 & -20 \\ 3 & -1 & -24 \\ 0 & 2 & -12 \end{bmatrix} x + \begin{bmatrix} 2 \\ 1 \\ 1 \end{bmatrix} u + \begin{bmatrix} 20 \\ 25 \\ 12 \end{bmatrix} y$$

(2) Using canonical form

Given an observable single-output system

$$\left.\begin{aligned} \dot{x}(t) &= Ax(t) + Bu(t) \\ y(t) &= Cx(t) \end{aligned}\right\} \tag{4.23}$$

and a set of desired poles, $\hat{s}_1, \hat{s}_2, \cdots, \hat{s}_n$, of the observer, the procedure of determining feedback gain matrix G is summarized as follows.

(1) Find the characteristic polynomial of A matrix

$$\det(sI - A) = s^n + \alpha_1 s^{n-1} + \cdots + \alpha_{n-1}s + \alpha_n \tag{4.24}$$

(2) Find the transformation matrix P_o and P_o^{-1} that transform (A, C) into the observable canonical form (A_o, C_o) by a suitable transformation $x = P_o z$. The transformation matrix is (see Eq. (3.23))

$$P_o^{-1} = \begin{bmatrix} a_{n-1} & a_{n-2} & \cdots & a_1 & 1 \\ a_{n-2} & a_{n-3} & \cdots & 1 & 0 \\ \vdots & \vdots & & \vdots & \vdots \\ a_1 & 1 & \cdots & 0 & 0 \\ 1 & 0 & \cdots & 0 & 0 \end{bmatrix} Q_o \tag{4.25}$$

(3) Calculate the desired characteristic polynomial

$$\Delta_d(s) = (s - \hat{s}_1)(s - \hat{s}_2)\cdots(s - \hat{s}_n) = s^n + \hat{\alpha}_1 s^{n-1} + \cdots + \hat{\alpha}_{n-1}s + \hat{\alpha}_n \tag{4.26}$$

(4) Find the feedback gain matrix with respect to the new state G_z. Denoting $G_z = [g_{z1} \quad g_{z2} \quad \cdots \quad g_{zn}]^T$ yields

$$|sI - (A_o - G_z C_o)| = \left| sI - \left[\begin{array}{c:c} \mathbf{0} & -(\alpha_n + g_{z1}) \\ & -(\alpha_{n-1} + g_{z2}) \\ \hdashline I_{n-1} & \vdots \\ & -(\alpha_1 + g_{zn}) \end{array}\right] \right| =$$

$$s^n + (\alpha_1 + g_{zn})s^{n-1} + \cdots + (\alpha_{n-1} + g_{z2})s + (\alpha_n + g_{z1})$$

Comparing the above polynomial with Eq. (4.26), we get

$$G_z = [g_{z1} \quad g_{z2} \quad \cdots \quad g_{zn}]^T = [\hat{\alpha}_n - \alpha_n \quad \hat{\alpha}_{n-1} - \alpha_{n-1} \quad \cdots \quad \hat{\alpha}_1 - \alpha_1]^T \tag{4.27}$$

(5) Determine required feedback gain matrix G. From the transformation $x = P_o z$ we have

$$\dot{z}_e = P_o^{-1}(A - GC)P_o z_e + P_o^{-1}Bu + P_o^{-1}Gy = (P_o^{-1}AP_o - P_o^{-1}GCP_o)z_e + P_o^{-1}Bu + P_o^{-1}Gy = (A_o - G_z C_o)z_e + P_o^{-1}Bu + P_o^{-1}Gy$$

Hence, from the above equation, we get the required feedback gain matrix is given by

$$G = P_o G_z \tag{4.28}$$

Example 4.4 Again consider the system given by

$$\dot{x} = \begin{bmatrix} 1 & 0 & 0 \\ 3 & -1 & 1 \\ 0 & 2 & 0 \end{bmatrix} x + \begin{bmatrix} 2 \\ 1 \\ 1 \end{bmatrix} u,$$

$$y = [0 \quad 0 \quad 1]x$$

Design an asymptotic observe such that its poles are located at $\hat{s}_1 = -3$, $\hat{s}_2 = -4$, and $\hat{s}_3 = -5$, via the method for the canonical form.

Solution It is know from Example 4.3 that the system is observable. The characteristic polynomial of the matrix A is

$$\Delta(s) = \det(sI - A) = s^3 - 3s + 2$$

The transformation matrix transforming (A, C) into the canonical form A_o, C_o is found as

$$P_o^{-1} = LQ_o = \begin{bmatrix} -3 & 0 & 1 \\ 0 & 1 & 0 \\ 1 & 0 & 0 \end{bmatrix}\begin{bmatrix} 0 & 0 & 1 \\ 0 & 2 & 0 \\ 6 & -2 & 2 \end{bmatrix} = \begin{bmatrix} 6 & -2 & -1 \\ 0 & 2 & 0 \\ 0 & 0 & 1 \end{bmatrix}$$

i.e.,

$$P_o = \frac{1}{12}\begin{bmatrix} 2 & 2 & 2 \\ 0 & 6 & 0 \\ 0 & 0 & 12 \end{bmatrix}$$

The required characteristic polynomial is

$$\Delta_d(s) = (s+3)(s+4)(s+5) = s^3 + 12s^2 + 47s + 60$$

Denoting $G_z = [g_{z1} \quad g_{z2} \quad g_{z3}]^T$, we get

$$G_z = [\hat{a}_3 - a_3 \quad \hat{a}_2 - a_2 \quad \hat{a}_1 - a_1]^T = [58 \quad 50 \quad 12]^T$$

Hence, the required feedback matrix is

$$G = P_o G_z = \frac{1}{12}\begin{bmatrix} 2 & 2 & 2 \\ 0 & 6 & 0 \\ 0 & 0 & 12 \end{bmatrix}\begin{bmatrix} 58 \\ 50 \\ 12 \end{bmatrix} = \begin{bmatrix} 20 \\ 25 \\ 12 \end{bmatrix}$$

and the state equation of the observer is

$$\dot{x}_e = (A - GC)x_e + Bu + Gy = \begin{bmatrix} 1 & 0 & -20 \\ 3 & -1 & -24 \\ 0 & 2 & -12 \end{bmatrix} x + \begin{bmatrix} 2 \\ 1 \\ 1 \end{bmatrix} u + \begin{bmatrix} 20 \\ 25 \\ 12 \end{bmatrix} y$$

3. Reduced-dimension Observer

The state observer presented above has the same dimension as the system to be estimated.

In fact, the dimension of the observer can actually less than this because the output already gives certain information of the state variables. Obviously, it is interested to design an observer with least possible dimension. In what follows we will show that if the rank of the output matrix is q, then an asymptotic observer of dimension $(n-q)$ can be designed to generate all the state variables. This will be achieved in two steps. First q state variables given directly from the output are separated. Then $(n-q)$ state variables are estimated by means of an observer.

Consider the n-dimensional system described by

$$\left.\begin{aligned}\dot{\boldsymbol{x}}(t) &= \boldsymbol{A}\boldsymbol{x}(t) + \boldsymbol{B}\boldsymbol{u}(t)\\ \boldsymbol{y}(t) &= \boldsymbol{C}\boldsymbol{x}(t)\end{aligned}\right\} \tag{4.29}$$

where $\boldsymbol{u}$ and $\boldsymbol{y}$ are $p \times 1$ and $q \times 1$ vectors respectively. It is assumed that the matrix $\boldsymbol{C}$ has rank q.

Construct a $n \times n$ transformation matrix

$$\boldsymbol{P} = \begin{bmatrix}\boldsymbol{R}\\ \boldsymbol{C}\end{bmatrix}^{-1} \tag{4.30}$$

where $(n-q) \times n$ real matrix $\boldsymbol{R}$ can be chosen arbitrarily such that $\boldsymbol{P}$ matrix is nonsingular. Introducing transformation $\boldsymbol{x} = \boldsymbol{P}\boldsymbol{z}$ leads

$$\boldsymbol{CP} = \boldsymbol{C}\begin{bmatrix}\boldsymbol{R}\\ \boldsymbol{C}\end{bmatrix}^{-1} = [\boldsymbol{0} \quad \boldsymbol{I}]\begin{bmatrix}\boldsymbol{R}\\ \boldsymbol{C}\end{bmatrix}\begin{bmatrix}\boldsymbol{R}\\ \boldsymbol{C}\end{bmatrix}^{-1} = [\boldsymbol{0} \quad \boldsymbol{I}] \tag{4.31}$$

and

$$\dot{\boldsymbol{z}} = \begin{bmatrix}\dot{\boldsymbol{z}}_1\\ \dot{\boldsymbol{z}}_2\end{bmatrix} = \begin{bmatrix}\bar{\boldsymbol{A}}_{11} & \bar{\boldsymbol{A}}_{12}\\ \bar{\boldsymbol{A}}_{21} & \bar{\boldsymbol{A}}_{22}\end{bmatrix}\begin{bmatrix}\boldsymbol{z}_1\\ \boldsymbol{z}_2\end{bmatrix} + \begin{bmatrix}\bar{\boldsymbol{B}}_1\\ \bar{\boldsymbol{B}}_2\end{bmatrix}\boldsymbol{u} \tag{4.32}$$

$$\boldsymbol{y} = [\boldsymbol{0} \quad \boldsymbol{I}]\begin{bmatrix}\boldsymbol{z}_1\\ \boldsymbol{z}_2\end{bmatrix} = \boldsymbol{z}_2 \tag{4.33}$$

Eq. (4.33) shows that only $(n-q)$ state variables $\boldsymbol{z}_1$ should be estimated by means of an observer, because q state variables $\boldsymbol{z}_2$ can be directly given from the output.

Substituting $\boldsymbol{z}_2 = \boldsymbol{y}$ into Eq. (4.32) yields

$$\left.\begin{aligned}\dot{\boldsymbol{z}}_1 &= \bar{\boldsymbol{A}}_{11}\boldsymbol{z}_1 + \bar{\boldsymbol{A}}_{12}\boldsymbol{z}_2 + \bar{\boldsymbol{B}}_1\boldsymbol{u} = \bar{\boldsymbol{A}}_{11}\boldsymbol{z}_1 + \bar{\boldsymbol{A}}_{12}\boldsymbol{y} + \bar{\boldsymbol{B}}_1\boldsymbol{u}\\ \dot{\boldsymbol{z}}_2 &= \bar{\boldsymbol{A}}_{21}\boldsymbol{z}_1 + \bar{\boldsymbol{A}}_{22}\boldsymbol{z}_2 + \bar{\boldsymbol{B}}_2\boldsymbol{u} = \bar{\boldsymbol{A}}_{21}\boldsymbol{z}_1 + \bar{\boldsymbol{A}}_{22}\boldsymbol{y} + \bar{\boldsymbol{B}}_2\boldsymbol{u} = \dot{\boldsymbol{y}}\end{aligned}\right\} \tag{4.34}$$

For the subsystem concerning the state variables $\boldsymbol{z}_1$ the vector $\boldsymbol{u}$ and $\boldsymbol{y}$ can be considered as external inputs. Thus Eq. (4.34) can be written as

$$\left.\begin{aligned}\dot{\boldsymbol{z}}_1 &= \bar{\boldsymbol{A}}_{11}\boldsymbol{z}_1 + \bar{\boldsymbol{u}}\\ \bar{\boldsymbol{y}} &= \bar{\boldsymbol{A}}_{21}\boldsymbol{z}_1\end{aligned}\right\} \tag{4.35}$$

where

$$\bar{\boldsymbol{u}} = \bar{\boldsymbol{A}}_{12}\boldsymbol{y} + \bar{\boldsymbol{B}}_1\boldsymbol{u} \tag{4.36}$$

$$\bar{\boldsymbol{y}} = \dot{\boldsymbol{y}} - \bar{\boldsymbol{A}}_{22}\boldsymbol{y} - \bar{\boldsymbol{B}}_2\boldsymbol{u} \tag{4.37}$$

If the system $(\boldsymbol{A}, \boldsymbol{B}, \boldsymbol{C})$ is observable, then the subsystem $(\bar{\boldsymbol{A}}_{11}, \bar{\boldsymbol{A}}_{21})$ must be observable. Thus a $(n-q)$-dimensional asymptotic observer can be obtained, from Eq (4.18), as

$$\dot{z}_1 = (\bar{A}_{11} - G_1\bar{A}_{21})z_1 + \bar{u} + G_1\bar{y} \tag{4.38}$$

Moreover, the poles of this observer can be assigned arbitrarily by choosing a suitable feedback gain matrix G_1.

Substituting Eq. (4.36) and Eq. (4.37) into Eq. (4.38) yields

$$\dot{z}_{1e} = (\bar{A}_{11} - G_1\bar{A}_{21})z_{1e} + (\bar{A}_{12}y + \bar{B}_1u) + G_1(\dot{y} - \bar{A}_{22}y - \bar{B}_2u) \tag{4.39}$$

it can be seen that this state equation contains $\dot{y}$. In order to avoid using differentiators in constructing the observer, a new state vector

$$w_e = z_{1e} - G_1y \tag{4.40}$$

is introduced. Thus, the state equation becomes

$$\begin{aligned}\dot{w}_e &= (\bar{A}_{11} - G_1\bar{A}_{21})(w_e + G_1y) + (\bar{A}_{12}y + \bar{B}_1u) + G_1(\dot{y} - \bar{A}_{22}y - \bar{B}_2u) - G_1\dot{y} = \\ &(\bar{A}_{11} - G_1\bar{A}_{21})w_e + (\bar{B}_1 - G_1\bar{B}_2)u + [(\bar{A}_{11} - G_1\bar{A}_{21})G_1 + \bar{A}_{12} - G_1\bar{A}_{22}]y\end{aligned} \tag{4.41}$$

Finally, the estimated state vector is given by

$$\hat{x}_e = Pz_e = P\begin{bmatrix} z_{1e} \\ y \end{bmatrix} = P\begin{bmatrix} w_e + G_1y \\ y \end{bmatrix} \tag{4.42}$$

as shown in Fig. 4.6.

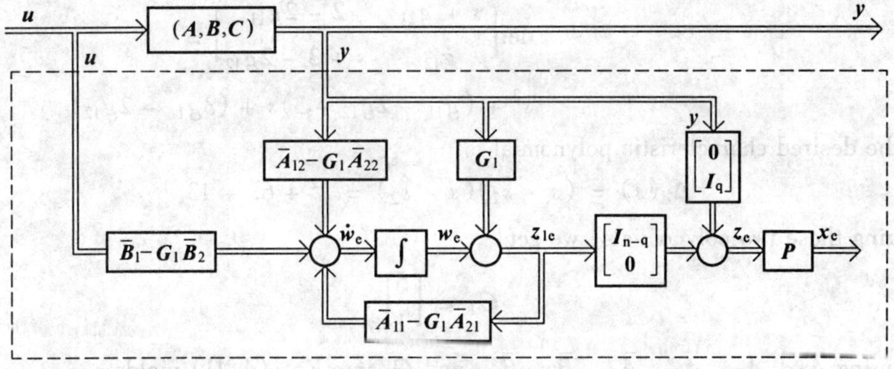

Figure 4.6 $(n - q)$-dimensional asymptotic state observer

Example 4.5 Consider the system given by

$$\dot{x} = \begin{bmatrix} 0 & 0 & 1 \\ 0 & -1 & 1 \\ 1 & 2 & -1 \end{bmatrix}x + \begin{bmatrix} 2 \\ 1 \\ 1 \end{bmatrix}u,$$

$$y = [0 \quad 2 \quad 1]x$$

Design an reduced-dimensional observer such that its poles are located at $\hat{s}_{1,2} = -3 \pm j2$.

Solution Checking the system's observability yields

$$\text{rank } Q_o = \text{rank}\begin{bmatrix} 0 & 2 & 1 \\ 1 & 0 & 1 \\ 1 & 2 & 0 \end{bmatrix} = 3$$

The system is observable, and a two-dimensional asymptotical state observer can be constructed.

First, we choose a nonsingular transformation matrix

$$P = \begin{bmatrix} R \\ C \end{bmatrix}^{-1} = \begin{bmatrix} 1 & 0 & 0 \\ 0 & 1 & 0 \\ 0 & 2 & 1 \end{bmatrix}^{-1} = \begin{bmatrix} 1 & 0 & 0 \\ 0 & 1 & 0 \\ 0 & -2 & 1 \end{bmatrix}$$

Introducing transformation $x = Pz$ yields

$$\dot{z} = \begin{bmatrix} \bar{A}_{11} & \bar{A}_{12} \\ \bar{A}_{21} & \bar{A}_{22} \end{bmatrix}\begin{bmatrix} z_1 \\ z_2 \end{bmatrix} + \begin{bmatrix} \bar{B}_1 \\ \bar{B}_2 \end{bmatrix} u = P^{-1}APz + P^{-1}Bu =$$

$$\left[\begin{array}{cc:c} 0 & -2 & 1 \\ 0 & -3 & 1 \\ \hdashline 1 & -2 & 1 \end{array}\right]\begin{bmatrix} z_1 \\ z_2 \end{bmatrix} + \left[\begin{array}{c} 2 \\ 1 \\ \hdashline 3 \end{array}\right] u$$

$$y = CPz = [0 \quad 0 \,\vdots\, 1]z$$

Since z_3 is directly observed, we construct an observer for z_1 and z_2.

The characteristic polynomial of the observer, from Eq. (4.38), is

$$\det(sI - (\bar{A}_{11} - G_1\bar{A}_{21})) = \det\left(\begin{bmatrix} s & \\ & s \end{bmatrix} - \begin{bmatrix} 0 & -2 \\ 0 & -3 \end{bmatrix} + \begin{bmatrix} g_{11} \\ g_{12} \end{bmatrix}[1 \quad -2]\right) =$$

$$\det\begin{bmatrix} s + g_{11} & 2 - 2g_{11} \\ g_{12} & s + 3 - 2g_{12} \end{bmatrix} =$$

$$s^2 + (g_{11} - 2g_{12} + 3)s + (3g_{11} - 2g_{12})$$

Since the desired characteristic polynomial is

$$\Delta_d(s) = (s - \hat{s}_1)(s - \hat{s}_2) = s^2 + 6s + 13$$

Comparing these two polynomials, we get

$$G_1 = \begin{bmatrix} 5 \\ 1 \end{bmatrix}$$

Substituting $\bar{A}_{11}$, $\bar{A}_{12}$, $\bar{A}_{21}$, $\bar{A}_{22}$, $\bar{B}_1$, $\bar{B}_2$, and G_1 into Eq. (4.41) yields

$$\dot{w}_e = (\bar{A}_{11} - G_1\bar{A}_{21})w_e + (\bar{B}_1 - G_1\bar{B}_2)u + [(\bar{A}_{11} - G_1\bar{A}_{21})G_1 + \bar{A}_{12} - G_1\bar{A}_{22}]y =$$

$$\left(\begin{bmatrix} 0 & -2 \\ 0 & -3 \end{bmatrix} - \begin{bmatrix} 5 \\ 1 \end{bmatrix}[1 \quad -2]\right) w_e + \left(\begin{bmatrix} 2 \\ 1 \end{bmatrix} - \begin{bmatrix} 5 \\ 1 \end{bmatrix} \cdot 3\right) u +$$

$$\left\{\left(\begin{bmatrix} 0 & -2 \\ 0 & -3 \end{bmatrix} - \begin{bmatrix} 5 \\ 1 \end{bmatrix}[1 \quad -2]\right)\begin{bmatrix} 5 \\ 1 \end{bmatrix} + \begin{bmatrix} 1 \\ 1 \end{bmatrix} - \begin{bmatrix} 5 \\ 1 \end{bmatrix} \cdot 1\right\} y =$$

$$\begin{bmatrix} -5 & 8 \\ -1 & -1 \end{bmatrix} w_e + \begin{bmatrix} -13 \\ -2 \end{bmatrix} u + \begin{bmatrix} -21 \\ -6 \end{bmatrix} y$$

Thus, from Eq. (4.40), we have

$$z_{1e} = \begin{bmatrix} z_{1e} \\ z_{2e} \end{bmatrix} = w_e + G_1 y = \begin{bmatrix} w_{1e} \\ w_{2e} \end{bmatrix} + \begin{bmatrix} 5 \\ 1 \end{bmatrix} y = \begin{bmatrix} w_{1e} + 5y \\ w_{2e} + y \end{bmatrix}$$

Finally, the estimated state vector is

$$x_e = Pz_e = \begin{bmatrix} 1 & 0 & 0 \\ 0 & 1 & 0 \\ 0 & -2 & 1 \end{bmatrix}\begin{bmatrix} z_{1e} \\ z_{2e} \\ z_{3e} \end{bmatrix} = \begin{bmatrix} 1 & 0 & 0 \\ 0 & 1 & 0 \\ 0 & -2 & 1 \end{bmatrix}\begin{bmatrix} w_{1e}+5y \\ w_{2e}+y \\ y \end{bmatrix} = \begin{bmatrix} w_{1e}+5y \\ w_{2e}+y \\ -2w_{2e}-y \end{bmatrix}$$

4.4.3 Separation Property

It is interested to us whether there is any difference in state feedback between using the estimated state $\boldsymbol{x}_e$ and the actual state $\boldsymbol{x}$.

Consider a controllable and observable n-dimensional system $(\boldsymbol{A}, \boldsymbol{B}, \boldsymbol{C})$, and then begin our discussion by setting up a combined system shown in Fig. 4.7, in which the compensator is consisted of an unreduced dimensional observer and a state feedback controller.

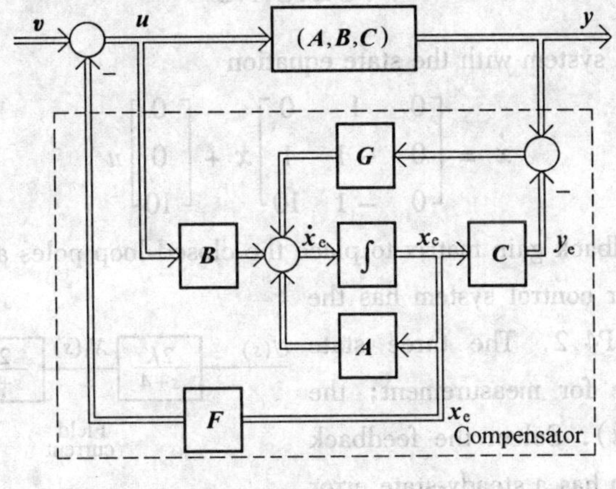

Figure 4.7 State feedback system with Unreduced dimensional observer

In this scheme, there are two real problems:

(a) The state feedback matrix $\boldsymbol{F}$ is designed with respect to the actual state, will it gives the same result if the estimated state is used in the feedback?

(b) Will the eigenvalues of the state observer appear in the entire system without any change?

From Fig. 4.7 the following equations can be written by inspection

$$\dot{\boldsymbol{x}} = \boldsymbol{A}\boldsymbol{x} - \boldsymbol{B}\boldsymbol{F}\boldsymbol{x}_e + \boldsymbol{B}\boldsymbol{v}$$

$$\dot{\boldsymbol{x}}_e = \boldsymbol{G}\boldsymbol{C}\boldsymbol{x} + (\boldsymbol{A} - \boldsymbol{G}\boldsymbol{C} - \boldsymbol{B}\boldsymbol{F})\boldsymbol{x}_e + \boldsymbol{B}\boldsymbol{v}$$

Furthermore, the state equation of the overall system can be rearranged in matrix form as

$$\begin{bmatrix} \dot{\boldsymbol{x}} \\ \dot{\boldsymbol{x}}_e \end{bmatrix} = \begin{bmatrix} \boldsymbol{A} & -\boldsymbol{B}\boldsymbol{F} \\ \boldsymbol{G}\boldsymbol{C} & \boldsymbol{A}-\boldsymbol{G}\boldsymbol{C}-\boldsymbol{B}\boldsymbol{F} \end{bmatrix}\begin{bmatrix} \boldsymbol{x} \\ \boldsymbol{x}_e \end{bmatrix} + \begin{bmatrix} \boldsymbol{B} \\ \boldsymbol{B} \end{bmatrix}\boldsymbol{v} \tag{4.43}$$

Introducing following nonsingular transformation

$$\begin{bmatrix} \boldsymbol{x} \\ \boldsymbol{x}_e \end{bmatrix} = \begin{bmatrix} \boldsymbol{I}_n & \boldsymbol{0} \\ \boldsymbol{I}_n & -\boldsymbol{I}_n \end{bmatrix}\begin{bmatrix} \boldsymbol{z} \\ \boldsymbol{z}_e \end{bmatrix}$$

Eq. (4.43) becomes

$$\begin{bmatrix} \dot{z} \\ \dot{z}_e \end{bmatrix} = \begin{bmatrix} A - BF & -BF \\ 0 & A - GC \end{bmatrix} \begin{bmatrix} z \\ z_e \end{bmatrix} + \begin{bmatrix} B \\ 0 \end{bmatrix} v \tag{4.44}$$

from Eq. (4.44) it can be seen immediately that the characteristic polynomial of the overall closed-loop system is

$$\Delta(s) = | sI - (A - BF) || sI - (A - GC) | \tag{4.45}$$

which means that as far as the eigenvalues are concerned, there is no difference in state feedback using the estimate state and the actual state. Consequently, the design of state feedback and the design of state observer can be carried out independently. This property is often called the separation property.

Problems

P4.1 Consider a system with the state equation

$$\dot{x} = \begin{bmatrix} 0 & 1 & 0 \\ 0 & -1 & 1 \\ 0 & -1 & 10 \end{bmatrix} x + \begin{bmatrix} 0 \\ 0 \\ 10 \end{bmatrix} u$$

And determine the feedback gain matrix to place the closed-loop poles at -10, $-1 \pm j\sqrt{3}$.

P4.2 A dc motor control system has the form shown in Fig. P4.2. The three state variables are available for measurement; the output position is $x_1(t)$. Select the feedback gains so that the system has a steady-state error equal to zero for a step input and a response with a percent overshoot less than 3%.

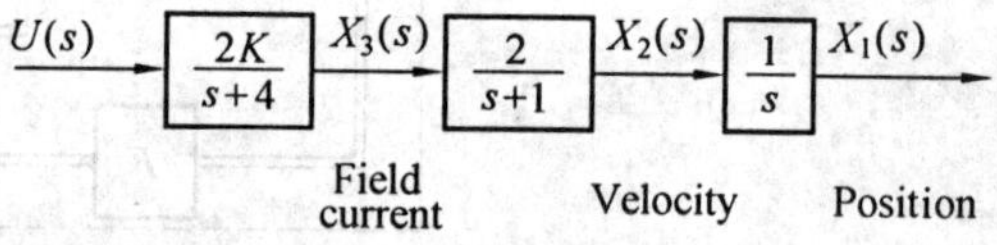

Figure P4.2 Field-controlled DC motor

P4.3 A control system has three physical state variables, as shown in Fig. P4.3. The state variable feedback structure is shown in the figure, with $f_1 = 1$.

(a) Select f_2 and f_3 so that the roots of the characteristic equation are three real roots lying between $s = -3$and $s = -6$.

(b) Select K_P so that the steady-state error for a step function is equal to zero.

(c) Plot the state variable diagram.

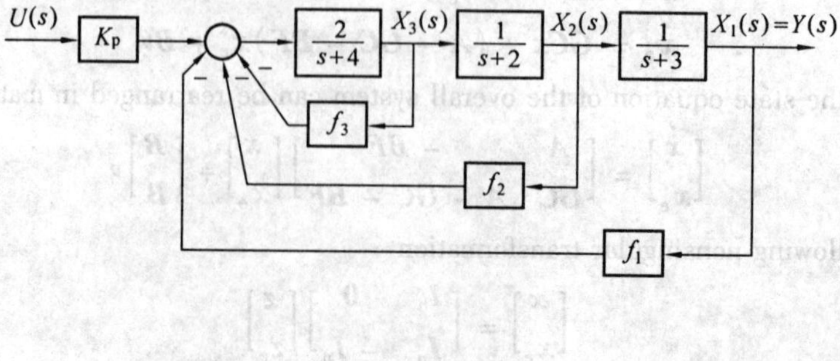

Figure P4.3 Automobile suspension system

P4.4 Consider a system with the transfer function

$$\frac{(s-1)(s+2)}{(s+1)(s-2)(s+3)}$$

Is it possible to change the transfer function to

$$\frac{(s-1)}{(s+2)(s+3)}$$

by state feedback. If yes, how?

P4.5 Design a state observer for the system represented by

$$\dot{x} = \begin{bmatrix} a & 1 \\ 0 & 0 \end{bmatrix} x + \begin{bmatrix} 0 \\ 1 \end{bmatrix} u,$$

$$y = [1 \quad 0] x$$

placing the observer eigenvalues at $-r$, $-2r(r > 0)$. Draw the state variable diagram.

P4.6 Consider a system with the state-space representation

$$\dot{x} = \begin{bmatrix} 1 & 2 & 0 \\ 3 & -1 & 1 \\ 0 & 2 & 0 \end{bmatrix} x + \begin{bmatrix} 0 \\ 0 \\ 1 \end{bmatrix} u,$$

$$y = [-1 \quad 1 \quad 1] x$$

Design a reduced dimensional state observer, selecting all the observer eigenvalues as -4.

P4.7 Consider a system with the state-space representation

$$\dot{x} = \begin{bmatrix} 0 & 1 & 4 \\ 3 & 2 & 0 \\ 1 & 1 & 1 \end{bmatrix} x + \begin{bmatrix} 0 & 0 \\ 1 & 0 \\ 0 & 1 \end{bmatrix} u,$$

$$y = \begin{bmatrix} 1 & 2 & 0 \\ 0 & 0 & 1 \end{bmatrix} x$$

(a) Design a reduced-dimensional state observer, placing the observer eigenvalue at -5.

(b) Design a state feedback controller so that the closed-loop poles are located at -1, -2, and -3.

(c) Draw the state variable diagram of the state feedback system with the reduced dimensional observer.

Chapter 5 Lyapunov Stability Analysis

Stability is an important qualitative property of control systems, because a workable system must be stable. It is easy to understand the concept of stability for the linear time-invariant systems in classical control theory. It is necessary now to extend the concept of stability to nonlinear systems and time-varying systems. Also, it is desired that the stability can be determined without explicitly solving for the eigenvalues. Lyapunov attacked this problem through his second, or direct, method. The sufficient but not necessary condition of this method requires the determination of a scalar function, called a Lyapunov function. The Lyapunov method provides a generalized approach to stability that is needed in the analysis and synthesis of modern control systems, such as the optimal control systems and self-adaptive control systems, which usually are nonlinear and/or time-varying.

5.1 Stability Definitions

5.1.1 Equilibrium State

The system which has no forcing input is represented by the state equation

$$\dot{\boldsymbol{x}} = \boldsymbol{f}(\boldsymbol{x}, t), \ \boldsymbol{x}(t_0) = \boldsymbol{x}_0 \tag{5.1}$$

where $\boldsymbol{x}$ is a $n \times 1$ state vector, t_0 is the initial time, and $\boldsymbol{f}(\boldsymbol{x}, t)$ is a linear or nonlinear, time-invariant or time-varying n-dimensional function, i.e.,

$$\dot{x}_i = f_i(x_1, x_2, \cdots, x_n; t), \ i = 1, 2, \cdots, n \tag{5.2}$$

In the study of stability, the solution to Eq. (5.1) is usually denoted by $\boldsymbol{x}(t; \boldsymbol{x}_0, t_0)$ to emphasize the initial state $\boldsymbol{x}_0$ and initial time t_0. Geometrically, the response $\boldsymbol{x}(t; \boldsymbol{x}_0, t_0)$ represents a trajectory released from the initial state $\boldsymbol{x}_0$ at initial time t_0 in a n-dimensional state space.

An important consideration in our stability analysis is that the response of a free-input system in the neighborhood of equilibrium state is closely related to the system's stability. A state $\boldsymbol{x}_e$ is said to be an equilibrium state if

$$\dot{\boldsymbol{x}}_e = \boldsymbol{f}(\boldsymbol{x}_e, t) = \boldsymbol{0} \tag{5.3}$$

for any t_0 and all $t \geqslant t_0$.

It can be seen, from this definition, that if a state trajectory in the state space reaches an equilibrium state and if no input is applied, the trajectory will stay at the equilibrium state forever. Consequently, an equilibrium state of $\dot{\boldsymbol{x}} = \boldsymbol{f}(\boldsymbol{x}, t)$ is a solution to $\dot{\boldsymbol{x}} = \boldsymbol{f}(\boldsymbol{x}, t) = \boldsymbol{0}$. Since the equilibrium state of a linear time-invariant system $\dot{\boldsymbol{x}} = \boldsymbol{A}\boldsymbol{x}$ satisfies $\boldsymbol{A}\boldsymbol{x}_e = \boldsymbol{0}$, clearly, there exists a unique zero solution $\boldsymbol{x} = \boldsymbol{0}$ if $\boldsymbol{A}$ is nonsingular, i.e., there exists an equilibrium state located at the origin of the state space. In this case, hence, the stability of a system is equivalent to the stability of the equilibrium state. As for a nonlinear system, the number of equilibrium state is depended on the state equation, because $\dot{\boldsymbol{x}} = \boldsymbol{f}(\boldsymbol{x}, t) = \boldsymbol{0}$ may have more than one solutions.

5.1.2 Stability in the Sense of Lyapunov

An equilibrium state $\boldsymbol{x}_e$ is said to be stable in the sense of Lyapunov if for any positive ε, there exists a positive number $\delta(\varepsilon, t_0)$, such that any $\| \boldsymbol{x}_0 - \boldsymbol{x}_e \| \leqslant \delta$ implies that

$$\| \boldsymbol{x}(t; \boldsymbol{x}_0, t_0) - \boldsymbol{x}_e \| \leqslant \varepsilon \tag{5.4}$$

for any t_0 and all $t \geqslant t_0$. In Eq. (5.4), symbol $\| \cdot \|$ is the norm of a vector. The norm of a vector is the generalization of the idea of length. Roughly speaking, it is a length measurement of a vector in the state space. Here, for example, $\| \boldsymbol{x}_0 - \boldsymbol{x}_e \|$ is used to represent the "length" from the point $\boldsymbol{x}_0$ to point $\boldsymbol{x}_e$ in the state space, and defined as

$$\| \boldsymbol{x}_0 - \boldsymbol{x}_e \| = \sqrt{(x_{10} - x_{1e})^2 + (x_{20} - x_{2e})^2 + \cdots + (x_{n0} - x_{ne})^2} \tag{5.5}$$

The stability in the sense of Lyapunov has an obvious geometric meaning. The expression $\| \boldsymbol{x}_0 - \boldsymbol{x}_e \| \leqslant \delta$ can be considered as a super-sphere region of δ radius and center at the equilibrium state $\boldsymbol{x}_e$ in the state space, and denoted by $S(\delta)$; similarly, the expression $\| \boldsymbol{x}(t; \boldsymbol{x}_0, t_0) - \boldsymbol{x}_e \| \leqslant \varepsilon$ is considered as a super-sphere region of ε radius and center at the equilibrium state $\boldsymbol{x}_e$ in the state space, and denoted by $S(\varepsilon)$. In this way, the stability in the sense of Lyapunov means that any trajectory starting in $S(\delta)$ does not leave $S(\varepsilon)$ for all $t \geqslant t_0$.

The above geometric meaning for a two-dimensional system is illustrated with a trajectory in the state plane as shown in Fig. 5.1(a). The trajectory ① is closed, which indicates a continuous oscillation, and trajectory ② terminates at the equilibrium state. In both two cases the equilibrium state $\boldsymbol{x}_e$ is stable in the sense of Lyapunov.

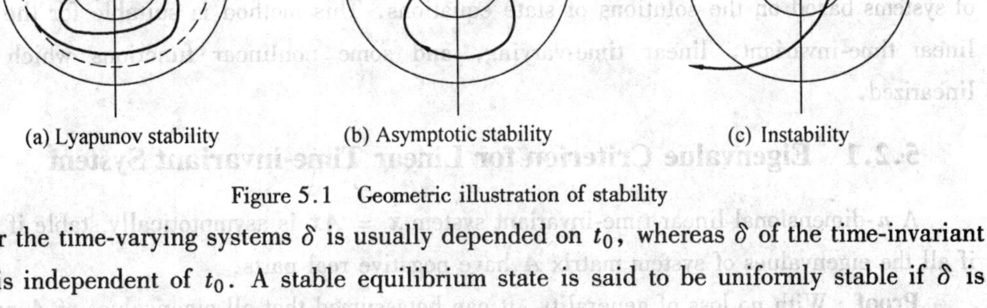

(a) Lyapunov stability (b) Asymptotic stability (c) Instability

Figure 5.1 Geometric illustration of stability

For the time-varying systems δ is usually depended on t_0, whereas δ of the time-invariant system is independent of t_0. A stable equilibrium state is said to be uniformly stable if δ is independent of t_0.

5.1.3 Asymptotic Stability

An equilibrium state $\boldsymbol{x}_e$ is said to be asymptotically stable if it is stable in the sense of Lyapunov and if

$$\lim_{t\to\infty} \| \boldsymbol{x}(t; x_0, t_0) - \boldsymbol{x}_e \| \to 0 \tag{5.6}$$

This means that all trajectories starting in $S(\delta)$ does not leave $S(\varepsilon)$ and approach the equilibrium state as time t approaches infinity. Obviously, this is consistent with the stability definition usually used in the classical control system. The geometric illustration of asymptotic stability is shown in Fig. 5.1(b).

An equilibrium state $\boldsymbol{x}_e$ is said to be uniformly asymptotically stable if δ is independent of t_0.

5.1.4 Globally Stability

A stable equilibrium state is said globally stable or stable in the large if the region $S(\delta)$ can be extended to the entire state space. If the equilibrium state of a linear system is stable, it must be globally stable, because any initial state yields the same stability determination. As for the nonlinear systems, usually, the stability is closely related to the initial condition and a stable equilibrium state is only stable in the small.

A globally stable equilibrium state $\boldsymbol{x}_e$ is said to be uniformly globally stable if δ is independent of t_0.

5.1.5 Instability

An equilibrium state is said unstable if there exists any trajectory starting in $S(\delta)$, which leaves $S(\varepsilon)$ no matter how large ε and how small δ are specified. The geometric illustration of instability is shown in Fig. 5.1(c). Obviously, a linear system will be unstable if its equilibrium state is unstable.

5.2 First Method of Lyapunov

The first method of Lyapunov, i.e. the indirect method, is used to determine the stability of systems based on the solutions of state equations. This method is suitable for the cases of linear time-invariant, linear time-varying, and some nonlinear functions which can be linearized.

5.2.1 Eigenvalue Criterion for Linear Time-invariant System

A n-dimensional linear time-invariant system $\dot{\boldsymbol{x}} = \boldsymbol{A}\boldsymbol{x}$ is asymptotically stable if and only if all the eigenvalues of system matrix $\boldsymbol{A}$ have negative real parts.

Proof With no loss of generality, it can be assumed that all eigenvalues of $\boldsymbol{A}$ matrix are distinct. Thus, $\boldsymbol{A}$ can be diagonalized by introducing a suitable nonsingular iinear transformation $\boldsymbol{x} = \boldsymbol{P}\boldsymbol{z}$, i.e.,

$$\boldsymbol{P}^{-1}\boldsymbol{A}\boldsymbol{P} = \boldsymbol{\Lambda} = \mathrm{diag}(\lambda_1, \lambda_2, \cdots, \lambda_n)$$

and

$$\mathrm{e}^{\boldsymbol{A}t} = \boldsymbol{P}\mathrm{e}^{\boldsymbol{P}^{-1}\boldsymbol{A}\boldsymbol{P}t}\boldsymbol{P}^{-1} = \boldsymbol{P}\mathrm{e}^{\boldsymbol{\Lambda}t}\boldsymbol{P}^{-1}$$

Expanding this equation, it can be seen that every element of $\mathrm{e}^{\boldsymbol{A}t}$ is a linear combination of $\mathrm{e}^{\lambda_1 t}$, $\mathrm{e}^{\lambda_2 t}, \cdots, \mathrm{e}^{\lambda_n t}$. Thus the solution of $\dot{\boldsymbol{x}} = \boldsymbol{A}\boldsymbol{x}$ can be represented as

$$\boldsymbol{x}(t) = \mathrm{e}^{\boldsymbol{A}t}\boldsymbol{x}(0) = \boldsymbol{P}\mathrm{e}^{\boldsymbol{\Lambda}t}\boldsymbol{P}^{-1}\boldsymbol{x}(0) = [\boldsymbol{R}_1\mathrm{e}^{\lambda_1 t} + \boldsymbol{R}_2\mathrm{e}^{\lambda_2 t} + \cdots + \boldsymbol{R}_n\mathrm{e}^{\lambda_n t}\boldsymbol{P}^{-1}]\boldsymbol{x}(0)$$

where $\boldsymbol{R}_i$, $i = 1, 2, \cdots, n$, are coefficient matrices. It is easy to see that the system is asymptotically stable, i.e. $\lim_{t\to\infty}\boldsymbol{x}(t) \to 0$ for any given $\boldsymbol{x}(0)$, if and only if $\mathrm{Re}(\lambda_i) < 0$, $i = 1, 2, \cdots, n$.

When the matrix $\boldsymbol{A}$ has repeated eigenvalues, the response modes, such as $t\mathrm{e}^{\lambda t}$, $t^2\mathrm{e}^{\lambda t}$, and etc. will be added into the expression of $\boldsymbol{x}(t)$. In this case, however, the stability conclusion is not changed.

As for the stability in the sense of Lyapunov, if $\boldsymbol{A}$ is transformed into the Jordan form, the sufficient and necessary condition is that all the eigenvalue of $\boldsymbol{A}$ have nonpositive (negative or zero) real parts and the order of every Jordan block associated with the eigenvalue with zero real part is 1.

Example 5.1 Consider the stability of two systems if their system matrices are given by

$$\text{(a) } \boldsymbol{A}_1 = \begin{bmatrix} 0 & 0 \\ 0 & 0 \end{bmatrix}, \quad \text{(b) } \boldsymbol{A}_2 = \begin{bmatrix} 0 & 1 \\ 0 & 0 \end{bmatrix}$$

respectively.

Solution (a) The system matrix is in the Jordan form, and both two eigenvalues are zeros. There are two Jordan blocks are associated $\lambda = 0$, and the order of both blocks is 1. Hence, the system is stable in the sense of Lyapunov, but it is not asymptotically stable.

(b) Also, the system matrix is in the Jordan form, and both two eigenvalues are zeros. Since the order of the Jordan block associated with $\lambda = 0$ is 2, the system is unstable.

In fact, it is easy to check these conclusions for these two systems by finding their responses. For case (a), the response is

$$\boldsymbol{x}(t) = \mathrm{e}^{\boldsymbol{A}_1 t}\boldsymbol{x}(0) = \begin{bmatrix} 1 & 0 \\ 0 & 1 \end{bmatrix}\begin{bmatrix} x_1(0) \\ x_2(0) \end{bmatrix} = \begin{bmatrix} x_1(0) \\ x_2(0) \end{bmatrix}$$

whereas the response in case (b) is

$$\boldsymbol{x}(t) = \mathrm{e}^{\boldsymbol{A}_2 t}\boldsymbol{x}(0) = \begin{bmatrix} 1 & t \\ 0 & 1 \end{bmatrix}\begin{bmatrix} x_1(0) \\ x_2(0) \end{bmatrix} = \begin{bmatrix} x_1(0) + tx_2(0) \\ x_2(0) \end{bmatrix}$$

Obviously, the first system is and only is stable in the sense of Lyapunov, and the second system is unstable.

5.2.2 Eigenvalue Criterion for Linearized Time-invariant System

Usually, a nonlinear system may have more than one equilibrium state. In the case that the nonlinear system can be expanded in a Taylor series about one equilibrium state, in a small neighborhood of the equilibrium state the system behaves like a linear system, and the eigenvalues of linearized system can be used to determine stability of the equilibrium state.

Assume that the nonlinear system $\dot{\boldsymbol{x}} = \boldsymbol{f}(\boldsymbol{x})$ can be expanded, about the equilibrium state $\boldsymbol{x}_e$, in the following Taylor series

$$\dot{\boldsymbol{x}} = \boldsymbol{f}(\boldsymbol{x}_e) + \left.\frac{\partial \boldsymbol{f}}{\partial \boldsymbol{x}^{\mathrm{T}}}\right|_{\boldsymbol{x}=\boldsymbol{x}_e} \cdot (\boldsymbol{x} - \boldsymbol{x}_e) + \boldsymbol{g}(\boldsymbol{x}) \tag{5.7}$$

where $\boldsymbol{g}(\boldsymbol{x})$ is the summation of the higher-order terms in the Taylor series. Denoting

$$\Delta \boldsymbol{x} = \boldsymbol{x} - \boldsymbol{x}_e \tag{5.8}$$

and neglecting the higher-order terms in the Taylor series yield the linearized state equation

$$\Delta \dot{\boldsymbol{x}} = \left.\frac{\partial \boldsymbol{f}}{\partial \boldsymbol{x}^{\mathrm{T}}}\right|_{\boldsymbol{x}=\boldsymbol{x}_e} \cdot \Delta \boldsymbol{x} = \begin{bmatrix} \frac{\partial f_1}{\partial x_1} & \frac{\partial f_1}{\partial x_2} & \cdots & \frac{\partial f_1}{\partial x_n} \\ \frac{\partial f_2}{\partial x_1} & \frac{\partial f_2}{\partial x_2} & \cdots & \frac{\partial f_2}{\partial x_n} \\ \vdots & \vdots & & \vdots \\ \frac{\partial f_n}{\partial x_1} & \frac{\partial f_n}{\partial x_2} & \cdots & \frac{\partial f_n}{\partial x_n} \end{bmatrix}_{\boldsymbol{x}=\boldsymbol{x}_e} \cdot \Delta \boldsymbol{x} = \boldsymbol{J}_x \Delta \boldsymbol{x} \tag{5.9}$$

where f_i is the ith row of $\boldsymbol{f}(\boldsymbol{x})$, $\boldsymbol{J}_x = \partial \boldsymbol{f}/\partial \boldsymbol{x}^{\mathrm{T}}$ is called the Jacobi matrix and is evaluated at $\boldsymbol{x}_e$.

For the nonlinear system linearized as Eq. (5.9), the equilibrium state $\boldsymbol{x}_e$ will be asymptotically stable if all the eigenvalues of $\boldsymbol{J}_x$ have negative real parts; and $\boldsymbol{x}_e$ will be unstable if any eigenvalue has positive real part. As for the case that some eigenvalues have zero real parts and others have negative real parts, the stability of the nonlinear system at $\boldsymbol{x}_e$ is related to $\boldsymbol{g}(\boldsymbol{x})$. If $\boldsymbol{g}(\boldsymbol{x}) = \boldsymbol{0}$ the equilibrium state $\boldsymbol{x}_e$ will be stable in the sense of Lyapunov or unstable as the case for linear system, otherwise the stability cannot be determined by the first method of Lyapunov.

5.3 Second Method of Lyapunov

It can be observed from the vibration phenomenon of mechanical systems, the system will approach the equilibrium state early or later if its energy, including kinetic energy and potential energy, is decreased as time is elapsed. Since it is difficult to find the real expression of energy function for a control system, Lyapunov proposed to construct a virtual energy function $V(\boldsymbol{x}, t)$, so called Lyapunov function, relating to the state variables. Consequently, the decay of energy can be characterized by $\dot{V}(\boldsymbol{x}, t)$. The second, or direct, method of Lyapunov determines the stability of a system using the signs, i.e. the definiteness, of $V(\boldsymbol{x}, t)$ and

$\dot{V}(\boldsymbol{x},t)$ without solving the eigenvalues. This method is applicable to linear or nonlinear, time-invariant or time-varying, and deterministic or stochastic systems. The principal difficulty in applying this method is in formulating a correct Lyapunov function because the failure of one function to meet the stability condition does not mean that a true Lyapunov function does not exist.

5.3.1 Definiteness of Scalar Function and Sylvester Theorem

1. Definiteness of Scalar Function

The sign definiteness for a scalar function of a vector, such as $V(\boldsymbol{x})$, is defined for a spherical region S about the origin described by $\|\boldsymbol{x}\| \leqslant K$ (a constant equal to the radius of S).

A scalar function $V(\boldsymbol{x})$ is called positive definite when $V(\boldsymbol{x}) = 0$ for $\boldsymbol{x} = \boldsymbol{0}$ and $V(\boldsymbol{x}) > 0$ for all other $\boldsymbol{x} \neq \boldsymbol{0}$. For example $V(\boldsymbol{x}) = x_1^2 + x_2^2$ is positive definite.

A scalar function $V(\boldsymbol{x})$ is called positive semidefinite when $V(\boldsymbol{x}) = 0$ for $\boldsymbol{x} = \boldsymbol{0}$ and $V(\boldsymbol{x}) \geqslant 0$ for all other $\boldsymbol{x} \neq \boldsymbol{0}$. For example $V(\boldsymbol{x}) = (x_1 + x_2)^2$ is positive semidefinite.

A scalar function $V(\boldsymbol{x})$ is called negative definite when $V(\boldsymbol{x}) = 0$ for $\boldsymbol{x} = \boldsymbol{0}$ and $V(\boldsymbol{x}) < 0$ for all other $\boldsymbol{x} \neq \boldsymbol{0}$. If $-V(\boldsymbol{x})$ is positive definite then $V(\boldsymbol{x})$ is negative definite. For example $V(\boldsymbol{x}) = -(x_1^2 + x_2^2)$ is negative definite.

A scalar function $V(\boldsymbol{x})$ is called negative semidefinite when $V(\boldsymbol{x}) = 0$ for $\boldsymbol{x} = \boldsymbol{0}$ and $V(\boldsymbol{x}) \leqslant 0$ for all other $\boldsymbol{x} \neq \boldsymbol{0}$. If $-V(\boldsymbol{x})$ is positive semidefinite then $V(\boldsymbol{x})$ is negative semidefinite. For example $V(\boldsymbol{x}) = -(x_1 + x_2)^2$ is negative semidefinite.

A scalar function $V(\boldsymbol{x})$ is called indefinite when it presents both positive and negative values for $\boldsymbol{x} \neq \boldsymbol{0}$. For example $V(\boldsymbol{x}) = x_1 x_2$ is indefinite.

For time-varying scalar function $V(\boldsymbol{x},t)$, the definition of definiteness is related to the time. For example, $V(\boldsymbol{x},t)$ is said to be positive definite in the region S if $V(\boldsymbol{x},t) = 0$ for $\boldsymbol{x} = \boldsymbol{0}$ and $V(\boldsymbol{x},t) > 0$ for all other $\boldsymbol{x} \neq \boldsymbol{0}$ in S for $t \geqslant t_0$.

2. Quadratic Form Function

The second method of Lyapunov used in determining stability of linear systems utilizes scalar functions expressed in quadratic form, which is one of the important scalar functions.

A quadratic form $V(\boldsymbol{x})$ is a real homogeneous polynomial in the real variables $x_1, x_2, \cdots, x_n$ of the form

$$V(\boldsymbol{x}) = \sum_{i=1}^{n}\sum_{j=1}^{n} p_{ij} x_i x_j \tag{5.10}$$

where all p_{ij} are real. This homogeneous polynomial can be denoted as

$$V(\boldsymbol{x}) = \boldsymbol{x}^{\mathrm{T}}\boldsymbol{P}\boldsymbol{x} = [\,x_1 \quad x_2 \quad \cdots \quad x_n\,]\begin{bmatrix} p_{11} & p_{12} & \cdots & p_{1n} \\ p_{21} & p_{22} & \cdots & p_{2n} \\ \vdots & \vdots & & \vdots \\ p_{n1} & p_{n2} & \cdots & p_{nn} \end{bmatrix}\begin{bmatrix} x_1 \\ x_2 \\ \vdots \\ x_n \end{bmatrix} \tag{5.11}$$

where $\boldsymbol{P}$ is a symmetric matrix. For example

$$V(\boldsymbol{x}) = x_1^2 - 4x_2^2 + 5x_3^2 + 6x_1x_2 - 10x_2x_3 =$$

$$\boldsymbol{x}^{\mathrm{T}}\boldsymbol{P}\boldsymbol{x} = [\,x_1 \quad x_2 \quad x_3\,]\begin{bmatrix} 1 & 3 & 0 \\ 3 & -4 & -5 \\ 0 & -5 & 5 \end{bmatrix}\begin{bmatrix} x_1 \\ x_2 \\ x_3 \end{bmatrix}$$

3. Sylvester Theorem

The definiteness of a quadratic form can be determined by inspection or by calculating the principal minors of $\boldsymbol{P}$ matrix, which is sometimes called the Sylvester theorem. A principal minor of $\boldsymbol{P}$ matrix is obtained by deleting any row(s) and the same numbered column(s). The leading principal minor is formed by the square array within $\boldsymbol{P}$ matrix that contains p_{11}. There are n leading principal minors for $n \times n$ matrix $\boldsymbol{P}$, which are given by

$$\Delta_1 = p_{11},\ \Delta_{12} = \begin{vmatrix} p_{11} & p_{12} \\ p_{21} & p_{22} \end{vmatrix},\ \Delta_{123} = \begin{vmatrix} p_{11} & p_{12} & p_{13} \\ p_{21} & p_{22} & p_{23} \\ p_{31} & p_{32} & p_{33} \end{vmatrix},\ \cdots,\ \Delta_{12\cdots n} = |\,\boldsymbol{P}\,|$$

The subscript of Δ are the rows and columns used to form the minor.

According to the Sylvester theorem, if all the leading principal minors of $\boldsymbol{P}$ are positive, then $V(\boldsymbol{x})$ is positive definite and $\boldsymbol{P}$ is also called positive definite; if $|\,\boldsymbol{P}\,| = 0$ and all the principal minors of $\boldsymbol{P}$ are nonnegative, then $V(\boldsymbol{x})$ is positive semidefinite and $\boldsymbol{P}$ is also called positive semidefinite; if $-\boldsymbol{P}$ is positive definite, or all the odd-order leading principal minors of $\boldsymbol{P}$ are negative and all the even-order leading principal minors of $\boldsymbol{P}$ are positive, then $V(\boldsymbol{x})$ is negative definite and $\boldsymbol{P}$ is also called negative definite; if $-\boldsymbol{P}$ is positive semidefinite then $V(\boldsymbol{x})$ is negative semidefinite and $\boldsymbol{P}$ is also called negative semidefinite; if $\boldsymbol{P}$ is not positive definite, positive semidefinite, negative definite, or negative semidefinite, then $V(\boldsymbol{x})$ is indefinite and $\boldsymbol{P}$ is also called indefinite.

Example 5.2 Determine the definiteness of

$$\boldsymbol{P} = \begin{bmatrix} 1 & 1 & 1 \\ 1 & 1 & 1 \\ 1 & 1 & 0 \end{bmatrix}$$

Solution The leading principal minors are evaluated to check for positive definiteness

$$\Delta_1 = 1,\ \Delta_{12} = \begin{vmatrix} 1 & 1 \\ 1 & 1 \end{vmatrix} = 0,\ \Delta_{123} = \begin{vmatrix} 1 & 1 & 1 \\ 1 & 1 & 1 \\ 1 & 1 & 0 \end{vmatrix} = 0$$

The condition for positive definiteness is not satisfied. Next the other principal minors are evaluated to check for definiteness

$$\Delta_{13} = \begin{vmatrix} 1 & 1 \\ 1 & 0 \end{vmatrix} = -1$$

Since the principal minors are not all nonnegative, $\boldsymbol{P}$ matrix is not positive semidefinite.

Similarly, $\boldsymbol{P}$ is not negative definite or negative semidefinite. Hence $\boldsymbol{P}$ is indefinite.

5.3.2 Lyapunov Stability Theory

Consider the nonlinear or linear system

$$\dot{\boldsymbol{x}} = f(\boldsymbol{x}, t) \tag{5.12}$$

With no loss of generality, it is assumed that the equilibrium state is at the origin of the state space, i.e., $\boldsymbol{x}_e = \boldsymbol{0}$.

1. Theorem for Asymptotic Stability

For the system described by Eq. (5.12), if there exists a scalar function $V(\boldsymbol{x}, t)$ having continuous first derivatives such that

(1) $V(\boldsymbol{x}, t)$ is positive definite;

(2) one of the following conditions is satisfied:

a. $\dot{V}(\boldsymbol{x}, t)$ is negative definite, or

b. $\dot{V}(\boldsymbol{x}, t)$ is negative semidefinite, and $\dot{V}[\boldsymbol{x}(t; \boldsymbol{x}_0, t_0)]$ does not keep zero for any $\boldsymbol{x} \neq \boldsymbol{0}$;

then the equilibrium state $\boldsymbol{x}_e = \boldsymbol{0}$ is asymptotically stable, and $V(\boldsymbol{x}, t)$ is said to be a Lyapunov function for the system of Eq. (5.12). Moreover, the equilibrium state $\boldsymbol{x}_e = \boldsymbol{0}$ is globally asymptotically stable if $V(\boldsymbol{x}, t) \to \infty$ as $\|\boldsymbol{x}\| \to \infty$.

This theorem can be illustrated with a second-order system. The function $V(\boldsymbol{x}) = x_1^2 + x_2^2$ is positive definite and is represented by the paraboloid surface as shown in Fig. 5.2(a). The value $V(\boldsymbol{x}) = c$ (a constant) is represented by the intersection of the surface $V(\boldsymbol{x})$ and the plane $z = c$. The projection of this intersection on the $[x_1, x_2]$-plane is a closed curve, an ellipse, around the origin. There is a family of such closed curves in the $[x_1, x_2]$-plane for different values of c. If $\dot{V}(\boldsymbol{x})$ is negative, the response trajectory starting from the initial state $\boldsymbol{x}_0$ crosses the ellipses for successively smaller values of $V(\boldsymbol{x})$ and finally approaches the origin as $t \to \infty$, as shown in Fig. 5.2(b). If $\dot{V}(\boldsymbol{x})$ is negative semidefinite, the response trajectory starting from the initial state $\boldsymbol{x}_0$ is tangential to a certain ellipse and $\dot{V}(\boldsymbol{x}) = 0$ at the tangent point. However, since $\dot{V}(\boldsymbol{x})$ does not keep zero for any $\boldsymbol{x} \neq \boldsymbol{0}$, the trajectory does not stay at the tangent point but moves continuously towards the origin, as shown in Fig. 5.2 (c).

Example 5.3 Consider the stability of the system given by

$$\dot{x}_1 = x_2 - x_1(x_1^2 + x_2^2)$$

$$\dot{x}_2 = -x_1 - x_2(x_1^2 + x_2^2)$$

Solution Letting $\dot{x}_1 = 0$ and $\dot{x}_2 = 0$ yields the solution $x_1 = 0$ and $x_2 = 0$, i.e. the origin is the unique equilibrium state of the system. A simple quadratic form scalar function which is positive definite is

$$V(\boldsymbol{x}) = x_1^2 + x_2^2$$

Its derivate is

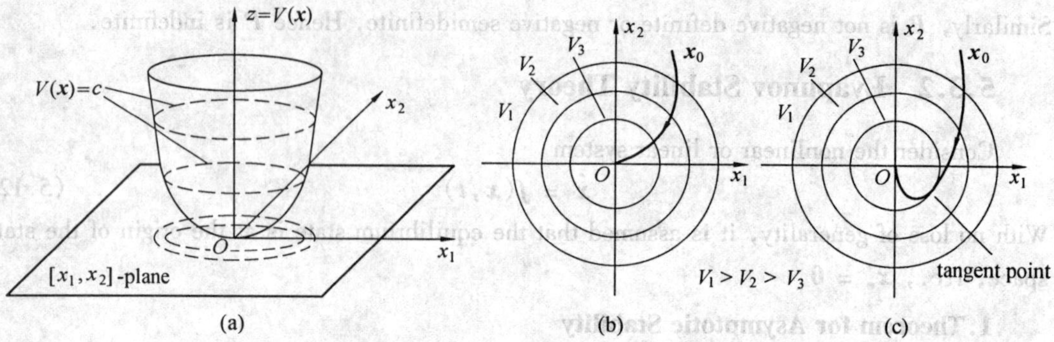

Figure 5.2 Positive definite function and projections on $[x_1, x_2]$-plane

$$\dot{V}(\boldsymbol{x}) = 2x_1\dot{x}_1 + 2x_2\dot{x}_2 = -2(x_1^2 + x_2^2)^2$$

The function $\dot{V}(\boldsymbol{x})$ is negative definite, which means that $V(\boldsymbol{x})$ is continuously decreased along any trajectory. Thus, $\boldsymbol{x}_e = \boldsymbol{0}$ is asymptotically stable, and the chosen $V(\boldsymbol{x})$ is a Liapunov function for the given system. Since $V(\boldsymbol{x}) \to \infty$ as $\|\boldsymbol{x}\| \to \infty$, $\boldsymbol{x}_e = \boldsymbol{0}$ is globally asymptotically stable.

Example 5.4 Consider the stability of the system given by

$$\dot{x}_1 = x_2$$
$$\dot{x}_2 = -x_1 - x_2$$

Solution Obviously, the origin $\boldsymbol{x} = \boldsymbol{0}$ is the unique equilibrium state of this system. Letting

$$V(\boldsymbol{x}) = 2x_1^2 + x_2^2$$

yields

$$\dot{V}(\boldsymbol{x}) = 4x_1\dot{x}_1 + 2x_2\dot{x}_2 = 2x_1x_2 - 2x_2^2$$

$V(\boldsymbol{x})$ is positive definite and $\dot{V}(\boldsymbol{x})$ is indefinite. It should be noted that this result does not mean that a true Lyapunov function does not exist. If we choose

$$V(\boldsymbol{x}) = x_1^2 + x_2^2$$

then

$$\dot{V}(\boldsymbol{x}) = 2x_1\dot{x}_1 + 2x_2\dot{x}_2 = -2x_2^2$$

Now, $V(\boldsymbol{x})$ is positive definite and $\dot{V}(\boldsymbol{x})$ is negative semidefinite. In this case it is necessary to investigate if $\dot{V}(\boldsymbol{x})$ keeps zero for any $\boldsymbol{x} \neq \boldsymbol{0}$. Letting $\dot{V}(\boldsymbol{x}) \equiv \boldsymbol{0}$ leads $x_2 \equiv 0$, which requires $\dot{x}_2 = 0$. From the system's equation, however, this requirement means that x_1 must be zero. Therefore, $\dot{V}(\boldsymbol{x})$ does not keep zero for any $\boldsymbol{x} \neq \boldsymbol{0}$. Thus, $\boldsymbol{x}_e = \boldsymbol{0}$ is asymptotically stable, and the chosen $V(\boldsymbol{x})$ is a Liapunov function. Moreover, since $V(\boldsymbol{x}) \to \infty$ as $\|\boldsymbol{x}\| \to \infty$, $\boldsymbol{x}_e = \boldsymbol{0}$ is globally asymptotically stable.

Also it should be noted that the Lyapunov function is not unique, for example $V(\boldsymbol{x}) = [(x_1 + x_2)^2 + 2x_1^2 + x_2^2]/2$ is another Lyapunov function for the system of Example 5.4.

2. Theorem for Stability in the Sense of Liapunov

For the system described by Eq. (5.12), $\boldsymbol{x}_e = \boldsymbol{0}$ is stable in the sense of Lyapunov if there exists a scalar function $V(\boldsymbol{x}, t)$ having continuous first derivatives such that

(1) $V(\boldsymbol{x}, t)$ is positive definite,

(2) $\dot{V}(\boldsymbol{x}, t)$ is negative semidefinite, and $\dot{V}[\boldsymbol{x}(t; \boldsymbol{x}_0, t_0)]$ keeps zero for some $\boldsymbol{x} \neq \boldsymbol{0}$, and $V(\boldsymbol{x}, t)$ is a Lyapunov function for the system of Eq. (5.12).

Example 5.5 Consider the stability of the following system

$$\dot{x}_1 = kx_2$$

$$\dot{x}_2 = -x_1$$

where $k > 0$.

Solution Obviously, the given system has a unique equilibrium state at the origin $\boldsymbol{x} = \boldsymbol{0}$. Choosing a positive scalar function

$$V(\boldsymbol{x}) = x_1^2 + kx_2^2$$

yields

$$\dot{V}(\boldsymbol{x}) = 2x_1\dot{x}_1 + 2kx_2\dot{x}_2 = 2kx_1x_2 - 2kx_1x_2 = 0$$

It is seen that $\dot{V}(\boldsymbol{x})$ will keep zero for any $\boldsymbol{x} \neq \boldsymbol{0}$. Hence the system is stable in the sense of Lyapunov, but not asymptotically stable.

3. Theorem for Instability

For the system described by Eq. (5.12), $\boldsymbol{x}_e = \boldsymbol{0}$ is unstable if there exists a scalar function $V(\boldsymbol{x}, t)$ having continuous first derivatives such that

(1) $V(\boldsymbol{x}, t)$ is positive definite,

(2) one of the following conditions is satisfied:

a. $\dot{V}(\boldsymbol{x}, t)$ is positive definite; or

b. $\dot{V}(\boldsymbol{x}, t)$ is positive semidefinite, and $\dot{V}[\boldsymbol{x}(t; \boldsymbol{x}_0, t_0)]$ does not keep zero for any $\boldsymbol{x} \neq \boldsymbol{0}$.

Example 5.6 Consider the stability of the following system

$$\dot{x}_1 = x_1 + x_2$$

$$\dot{x}_2 = -x_1 + x_2$$

Solution The given system has a unique equilibrium state at the origin $\boldsymbol{x} = \boldsymbol{0}$. Choosing a positive scalar function

$$V(\boldsymbol{x}) = x_1^2 + x_2^2$$

yields

$$\dot{V}(\boldsymbol{x}) = 2x_1\dot{x}_1 + 2x_2\dot{x}_2 = 2x_1^2 + 2x_2^2$$

By inspection, $\dot{V}(\boldsymbol{x})$ is positive definite and $\boldsymbol{x}_e = \boldsymbol{0}$ is unstable.

Example 5.7 Consider the stability of the following system

$$\dot{x}_1 = x_2$$

$$\dot{x}_2 = -x_1 + x_2$$

Solution Also, the given system has a unique equilibrium state at the origin $\boldsymbol{x} = \boldsymbol{0}$. Choosing a positive scalar function

$$V(\boldsymbol{x}) = x_1^2 + x_2^2$$

yields

$$\dot{V}(\boldsymbol{x}) = 2x_1\dot{x}_1 + 2x_2\dot{x}_2 = 2x_2^2$$

It is seen that $\dot{V}(\boldsymbol{x})$ is negative semidefinite. Letting $\dot{V}(\boldsymbol{x}) \equiv \boldsymbol{0}$ leads $x_2 \equiv 0$, which requires $\dot{x}_2 = 0$. From the system's state equation, however, this requirement means that x_1 must be zero. This means that $\dot{V}(\boldsymbol{x})$ does not keep zero for any $\boldsymbol{x} \neq \boldsymbol{0}$. Therefore $\boldsymbol{x}_e = \boldsymbol{0}$ is unstable.

5.4 Application of the Lyapunov Method to Linear Systems

The second method of Lyapunov is applicable to time-varying and nonlinear systems. Methods have been developed for many such systems, although there is no simple general method of developing the Lyapunov function $V(\boldsymbol{x}, t)$. Especially, the Lyapunov function $V(\boldsymbol{x}, t)$ for linear systems can be determined using following approach.

5.4.1 Time-invariant continuous system

Consider the n-dimensional unexcited system represented by the state equation

$$\dot{\boldsymbol{x}} = \boldsymbol{A}\boldsymbol{x} \tag{5.13}$$

where $\boldsymbol{A}$ is nonsingular. Obviously, the system has a unique state equilibrium at the origin. The possible Lyapunov function $V(\boldsymbol{x})$ is expressed in terms of the symmetric matrix $\boldsymbol{P}$ by

$$V(\boldsymbol{x}) = \boldsymbol{x}^{\mathrm{T}}\boldsymbol{P}\boldsymbol{x} \tag{5.14}$$

The derivative of $V(\boldsymbol{x})$ is

$$\dot{V}(\boldsymbol{x}) = \dot{\boldsymbol{x}}^{\mathrm{T}}\boldsymbol{P}\boldsymbol{x} + \boldsymbol{x}^{\mathrm{T}}\boldsymbol{P}\dot{\boldsymbol{x}} = \boldsymbol{x}^{\mathrm{T}}(\boldsymbol{A}^{\mathrm{T}}\boldsymbol{P} + \boldsymbol{P}\boldsymbol{A})\boldsymbol{x} \tag{5.15}$$

Letting

$$\boldsymbol{A}^{\mathrm{T}}\boldsymbol{P} + \boldsymbol{P}\boldsymbol{A} = -\boldsymbol{Q} \tag{5.16}$$

yields

$$\dot{V}(\boldsymbol{x}) = -\boldsymbol{x}^{\mathrm{T}}\boldsymbol{Q}\boldsymbol{x} \tag{5.17}$$

Eq. (5.16) is called the Lyapunov equation. It can be proven that, for the system given by Eq. (5.13), the state equilibrium at origin is asymptotically stable if and only if there exists a unique positive definite real symmetric matrix $\boldsymbol{P}$ which is satisfied the Lyapunov equation in Eq. (5.16) for any given positive definite real symmetric matrix $\boldsymbol{Q}$. Moreover, $\boldsymbol{x}^{\mathrm{T}}\boldsymbol{P}\boldsymbol{x}$ just is a Lyapunov function of the system.

From this criterion, the stability can be determined as following procedure.

Step 1. Choose an arbitrary $n \times n$ symmetric $\boldsymbol{Q}$ which is positive definite or positive semidefinite. For example, for convenience, a simple diagonal matrix such as $\boldsymbol{Q} = \boldsymbol{I}$ is positive definite. A positive semidefinite matrix $\boldsymbol{Q}$ can be chosen which contains all zero elements except one positive element in any position along the principal diagonal.

Step 2. Determine the elements of $\boldsymbol{P}$ by equating terms in the Lyapunov equation.

Step 3. Use the Sylvester theorem to determine the definiteness of $\boldsymbol{P}$.

Step 4. If $\boldsymbol{P}$ is positive definite then the state equilibrium at origin is asymptotically stable; if $\boldsymbol{P}$ is negative definite then the state equilibrium at origin is unstable.

Example 5.8 Determine stability for the system

$$\dot{\boldsymbol{x}} = \begin{bmatrix} 0 & 1 \\ -2 & -3 \end{bmatrix} \boldsymbol{x}$$

by using the Lyapunov equation.

Solution The symmetric matrix $\boldsymbol{P}$ is denoted as

$$\boldsymbol{P} = \begin{bmatrix} p_{11} & p_{12} \\ p_{12} & p_{22} \end{bmatrix}$$

and the matrix is chosen as $\boldsymbol{Q} = \boldsymbol{I}$. From the Lyapunov equation, we have

$$\begin{bmatrix} 0 & -2 \\ 1 & -3 \end{bmatrix} \begin{bmatrix} p_{11} & p_{12} \\ p_{12} & p_{22} \end{bmatrix} + \begin{bmatrix} p_{11} & p_{12} \\ p_{12} & p_{22} \end{bmatrix} \begin{bmatrix} 0 & 1 \\ -2 & -3 \end{bmatrix} = \begin{bmatrix} -1 & 0 \\ 0 & -1 \end{bmatrix}$$

From this identity we get three simultaneous equations

$$\begin{cases} -4p_{12} = 0 \\ -2p_{22} + p_{11} - 3p_{12} = 0 \\ 2p_{12} - 6p_{22} = -1 \end{cases}$$

The solution matrix is

$$\boldsymbol{P} = \begin{bmatrix} 5/4 & 1/4 \\ 1/4 & 1/4 \end{bmatrix}$$

Since $\boldsymbol{P}$ matrix is positive definite, the system is asymptotically stable. Moreover, a Lyapunov function can be obtained as

$$V(\boldsymbol{x}) = \boldsymbol{X}^{\mathrm{T}}\boldsymbol{P}\boldsymbol{X} = \frac{5}{4}x_1^2 + \frac{1}{2}x_1x_2 + \frac{1}{4}x_2^2$$

Example 5.9 Determine the range of values of k so that the system shown in Fig. 5.3 is asymptotically stable.

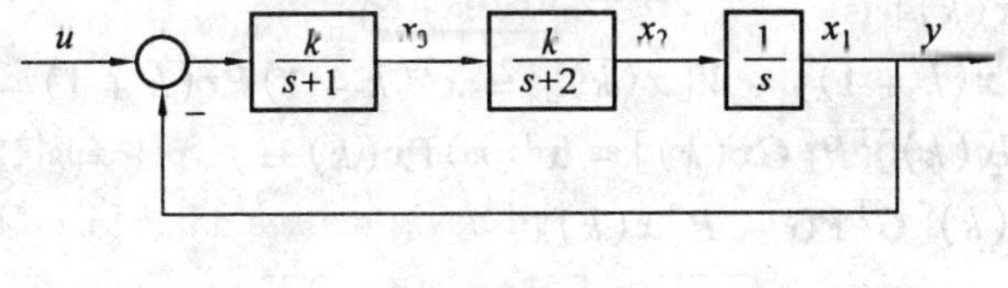

Figure 5.3 Block diagram for Example 5.9

Solution Assuming that the state variables are specified as shown in the block diagram, the state equation is given by

$$\dot{\boldsymbol{x}} = \begin{bmatrix} 0 & 1 & 0 \\ 0 & -2 & 1 \\ -k & 0 & -1 \end{bmatrix} \boldsymbol{x} + \begin{bmatrix} 0 \\ 0 \\ k \end{bmatrix} u$$

Letting $u = 0$ for studying the stability, since $\det A = -k \neq 0$, the system matrix is nonsingular and the origin is the unique equilibrium state of the autonomous system. If the matrix is chosen as positive semidefinite

$$Q = \begin{bmatrix} 0 & 0 & 0 \\ 0 & 0 & 0 \\ 0 & 0 & 1 \end{bmatrix}$$

then $\dot{V}(\boldsymbol{x}) = -x_3^2$ is negative semidefinite. Letting $\dot{V}(\boldsymbol{x}) \equiv 0$ yields $x_3 \equiv 0$. Consequently, taking into consideration of the state equation, we get $x_1 \equiv 0$ from $\dot{x}_3 = -kx_1 + x_3$ and $x_2 \equiv 0$ from $\dot{x}_1 = x_2$. This means that this positive semidefinite matrix $\boldsymbol{Q}$ can be used to simplify the analysis of stability.

The Lyapunov equation is

$$\begin{bmatrix} 0 & 0 & -k \\ 1 & -2 & 0 \\ 0 & 1 & -1 \end{bmatrix}\begin{bmatrix} p_{11} & p_{12} & p_{13} \\ p_{12} & p_{22} & p_{23} \\ p_{13} & p_{23} & p_{33} \end{bmatrix} + \begin{bmatrix} p_{11} & p_{12} & p_{13} \\ p_{12} & p_{22} & p_{23} \\ p_{13} & p_{23} & p_{33} \end{bmatrix}\begin{bmatrix} 0 & 1 & 0 \\ 0 & -2 & 1 \\ -k & 0 & -1 \end{bmatrix} = \begin{bmatrix} 0 & 0 & 0 \\ 0 & 0 & 0 \\ 0 & 0 & -1 \end{bmatrix}$$

and the resultant solution is

$$\boldsymbol{P} = \frac{1}{12-2k}\begin{bmatrix} k^2+12k & 6k & 0 \\ 6k & 3k & k \\ 0 & k & 6 \end{bmatrix}$$

From the Sylvester theorem, $\boldsymbol{P}$ is positive definite if and only if $0 < k < 6$. Hence, the system is asymptotically stable when $0 < k < 6$.

5.4.2 Time-invariant Discrete System

Consider the time-invariant discrete system represented by the state equation

$$\boldsymbol{x}(k+1) = \boldsymbol{G}\boldsymbol{x}(k) \tag{5.18}$$

where $\boldsymbol{G}$ is nonsingular. The system has a unique state equilibrium at the origin. Choose the possible Lyapunov function $V[\boldsymbol{x}(k)]$ is expressed in terms of the symmetric matrix $\boldsymbol{P}$ by

$$V[\boldsymbol{x}(k)] = \boldsymbol{x}^{\mathrm{T}}(k)\boldsymbol{P}\boldsymbol{x}(k) \tag{5.19}$$

The deference of $V[\boldsymbol{x}(k)]$ is

$$\begin{aligned}\Delta V[\boldsymbol{x}(k)] &= V[\boldsymbol{x}(k+1)] - V[\boldsymbol{x}(k)] = \boldsymbol{x}^{\mathrm{T}}(k+1)\boldsymbol{P}\boldsymbol{x}(k+1) - \boldsymbol{x}^{\mathrm{T}}(k)\boldsymbol{P}\boldsymbol{x}(k) = \\ &[\boldsymbol{G}\boldsymbol{x}(k)]^{\mathrm{T}}\boldsymbol{P}[\boldsymbol{G}\boldsymbol{x}(k)] - \boldsymbol{x}^{\mathrm{T}}(k)\boldsymbol{P}\boldsymbol{x}(k) = \\ &\boldsymbol{x}^{\mathrm{T}}(k)[\boldsymbol{G}^{\mathrm{T}}\boldsymbol{P}\boldsymbol{G} - \boldsymbol{P}]\boldsymbol{x}(k)\end{aligned} \tag{5.20}$$

Letting

$$\boldsymbol{G}^{\mathrm{T}}\boldsymbol{P}\boldsymbol{G} - \boldsymbol{P} = -\boldsymbol{Q} \tag{5.21}$$

yields

$$\Delta V[\boldsymbol{x}(k)] = -\boldsymbol{x}^{\mathrm{T}}(k)\boldsymbol{Q}\boldsymbol{x}(k) \tag{5.22}$$

Eq. (5.21) is called the discrete Lyapunov equation.

It can be proven that the discrete system given by Eq. (5.18) is asymptotically stable if and only if there exists a unique positive definite real symmetric matrix $\boldsymbol{P}$ which is satisfied the Lyapunov equation in Eq. (5.21) for any given positive definite real symmetric matrix $\boldsymbol{Q}$.

Problems

P5.1 Use the Jacobi matrix to determine the stability in the vicinity of the equilibrium points for the systems described by

(a) $\begin{cases} \dot{x}_1 = x_1 + x_1^2 - x_1 x_2 \\ \dot{x}_2 = x_1 - x_2 \end{cases}$

(b) $\begin{cases} \dot{x}_1 = 4x_1 - x_1 x_2 \\ \dot{x}_2 = x_1^2 - x_2 \end{cases}$

(c) $\begin{cases} \dot{x}_1 = -x_1^2 + x_2 \\ \dot{x}_2 = 4x_1 - 2x_2 \end{cases}$

P5.2 For the systems described by

(1) $\begin{cases} \dot{x}_1 = x_2 \\ \dot{x}_2 = -x_1 - 3x_2 + 0.25x_3^2 \\ \dot{x}_3 = x_1 x_3 - x_3 \end{cases}$

(2) $\begin{cases} \dot{x}_1 = x_2 \\ \dot{x}_2 = -x_2 - 2\sin x_1 \end{cases}$

(a) Determine the equilibrium points for each system.

(b) Obtain the linearized equations about the equilibrium points.

(c) Determine whether the system is asymptotically stable or unstable in the neighborhood of its equilibrium points.

P5.3 Write the following quadric functions in the vector-matrix form and determine their definiteness using the Sylvester criterion.

(a) $V(\boldsymbol{x}) = x_1^2 + 2x_2^2$

(b) $V(\boldsymbol{x}) = -x_1^2 - (3x_1^2 + 2x_2^2)^2$

(c) $V(\boldsymbol{x}) = -(x_1 + 2x_2)^2$

(d) $V(\boldsymbol{x}) = (x_1 + x_2)^2$

(e) $V(\boldsymbol{x}) = x_1 x_2 + x_2^2$

(f) $V(\boldsymbol{x}) = -x_1^2 - 3x_2^2 - 11x_3^2 + 2x_1 x_2 - 4x_2 x_3 - 2x_1 x_3$

(g) $V(\boldsymbol{x}) = x_1^2 + 4x_2^2 + x_3^2 + 2x_1 x_2 - 6x_2 x_3 - 2x_1 x_3$

P5.4 Use the Lyapunov second method to determine the stability for

(a) $\dot{\boldsymbol{x}} = \begin{bmatrix} -1 & 1 \\ 2 & -3 \end{bmatrix} \boldsymbol{x}$

(b) $\dot{\boldsymbol{x}} = \begin{bmatrix} 2 & 0.5 & -3 \\ 0 & -1 & 0 \\ 0 & 0.5 & -1 \end{bmatrix}$

P5.5 Use the Lyapunov equation to determine the range of K that yields a positive definite matrix $\boldsymbol{P}$ in the Lyapunov function $V(\boldsymbol{x}) = \boldsymbol{x}^{\mathrm{T}} \boldsymbol{P} \boldsymbol{x}$, thus assuring asymptotic stability for

(a) $A = \begin{bmatrix} -1 & K \\ 0 & -2 \end{bmatrix}$, $Q = \begin{bmatrix} 0 & 0 \\ 0 & 1 \end{bmatrix}$

(b) $A = \begin{bmatrix} -1 & 0 & 1 \\ 2 & -2 & 0 \\ K & 0 & -2 \end{bmatrix}$, $Q = \begin{bmatrix} 0 & 0 & 0 \\ 0 & 1 & 0 \\ 0 & 0 & 0 \end{bmatrix}$

P5.6 Use the Lyapunov equation to determine the range of K that yields a positive definite matrix $\boldsymbol{P}$ in the Lyapunov function $V[\boldsymbol{x}(k)] = \boldsymbol{x}^{\mathrm{T}}(k)\boldsymbol{P}\boldsymbol{x}(k)$, thus assuring asymptotic stability for

$$A = \begin{bmatrix} 0 & 1 & 0 \\ 0 & 0 & 1 \\ 0 & \frac{K}{2} & 10 \end{bmatrix}, \quad Q = \begin{bmatrix} 0 & 0 & 0 \\ 0 & 0 & 0 \\ 0 & 0 & 1 \end{bmatrix}$$

References

[1] 多尔夫,等.现代控制系统(英文影印版)[M].北京:科学出版社,2002.

[2] 德赖斯.线性控制系统工程(英文影印版)[M].北京:清华大学出版社,2002.

[3] J'AZZO J, H HOUPIS C. Linear Control System Analysis and Design [M]. New York: McGraw-Hill, 1988.

[4] 郑大钟.线性系统理论[M].北京:清华大学出版社,2002.

[5] 李友善.自动控制原理[M].北京:国防工业出版社,1989.

[6] 李文秀.自动控制原理[M].哈尔滨:哈尔滨工程大学出版社,2001.

[7] 胡寿松.自动控制原理[M].北京:科学出版社,2002.

[8] 陈启宗.线性系统理论与设计[M].北京:科学出版社,1988.

[9] 王诗宓,等.自动控制理论例题习题集[M].北京:清华大学出版社,2002.

第 1 章　线性系统的状态空间描述

1.1 引　言

现今对于控制工程师富有挑战意义的问题之一是诸如化工过程、机器人系统和交通管理系统之类大规模的、内部相互联系的现代复杂系统的建模和控制。尽管在经典控制理论中物理系统的传递函数描述提供了分析和设计的实用方法，而且也允许使用方框图表示子系统的内部连接，这种描述方法存在一些根本的局限。这是由于传递函数描述是控制系统的一种基于输入-输出关系的外部描述，它仅适用于线性时不变的单变量系统，即只有一个输入和一个输出的系统。

另一方面，基于状态空间描述的时域法则功能强多了。这种方法的主要优点是很容易在数字计算机上实施，适用于非线性系统和时变系统，很容易推广到多变量系统即具有若干个输入和输出的系统，它还能提供系统内部的行为表现，这些是传递函数分析法做不到的。

1.1.1　状态变量

状态和状态变量的概念是现代控制理论的基础，正如状态这个词的含义一样，一个系统的状态涉及该系统的过去、现在以及未来的状况。在一般情况下，状态可以用一组数字、一条曲线、一个方程或者其他什么更为抽象的东西描述。不过，对于动态系统而言，一个系统的状态则用一组状态变量来描述。

动态系统的状态变量定义为这样一组数量最少的变量，由 t_0 时这些变量的初始值和 $t \geqslant t_0$ 时任何作用于系统的输入 $u(t)$ 可以唯一地确定 $t \geqslant t_0$ 时该系统的行为。但是应当强调，状态变量的选择不是唯一的，对于一个给定的系统存在许多组不同的状态变量可供使用。

动态系统状态变量描述的一个简单例子是图 1.1 中所示的 RLC 电路。对于一个无源 RLC 网络，需要的状态变量数等于独立的储能元件数。这一系统的状态可以用一组状态变量 x_1、x_2 描述，其中 x_1 是电感器电流 $i_L(t)$，x_2 是电容器电压 $v_C(t)$。状态变量的这一选择直观上满足要求，因为该网络储存的能量可以用这些变量描述为

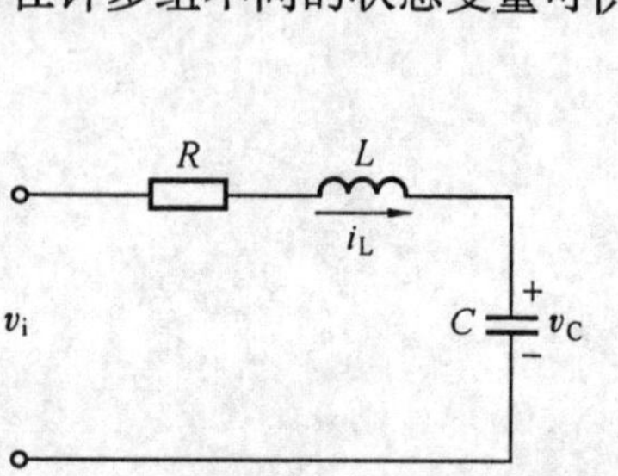

图 1.1　RLC 网络的状态变量模型

$$E = \frac{1}{2} L i_L^2 + \frac{1}{2} C v_C^2$$

应用基尔霍夫电流定律可以获得以下分别描述电容器电压和电感器电流变化速率的方程

$$L\frac{\mathrm{d}i_{\mathrm{L}}}{\mathrm{d}t} = -Ri_{\mathrm{L}} - v_{\mathrm{C}} + v_{\mathrm{i}}$$

$$C\frac{\mathrm{d}v_{\mathrm{C}}}{\mathrm{d}t} = i_{\mathrm{L}}$$

这些方程可以用状态变量 x_1 和 x_2 改写为一阶微分方程组如下

$$\frac{\mathrm{d}x_1}{\mathrm{d}t} = -\frac{R}{L}x_1 - \frac{1}{L}x_2 + \frac{1}{L}v_{\mathrm{i}}$$

$$\frac{\mathrm{d}x_2}{\mathrm{d}t} = \frac{1}{C}x_1$$

利用上述方程以及用 $x_1(t_0)$和 $x_2(t_0)$表示的该网络的初始条件，可以确定该系统未来的行为。

对于一个特定的系统，状态变量的数量是固定不变的，而且等于该系统的阶数。也可以说，状态变量的数量可以确定为求解该系统的微分方程时所需的初始条件数或者详细描述该系统时所需的一阶微分方程数。

1.1.2 状态空间表示

动态系统的一般形式如图 1.2 所示。

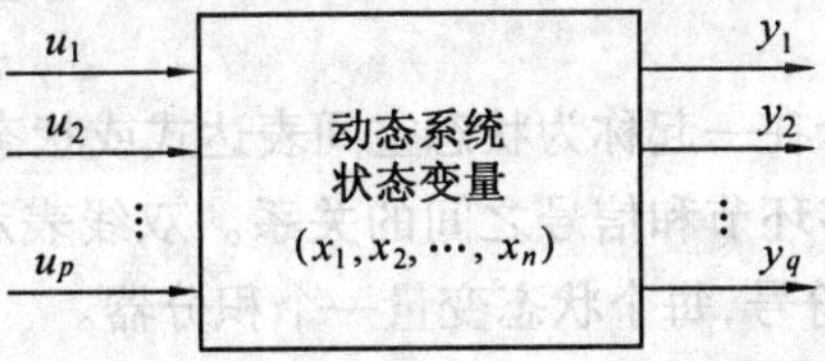

图 1.2　用状态变量描述的动态系统

由于系统的行为可以用状态变量完全描述，一个 n 阶线性时不变系统的状态可以用一组 n 个微分方程描述，用状态变量 $x_1, x_2, \cdots, x_n$ 写成的一般形式为

$$\left.\begin{aligned}
\dot{x}_1(t) &= a_{11}x_1(t) + a_{12}x_2(t) + \cdots + a_{1n}x_n(t) + b_{11}u_1(t) + \cdots + b_{1p}u_p(t)\\
\dot{x}_2(t) &= a_{21}x_1(t) + a_{22}x_2(t) + \cdots + a_{2n}x_n(t) + b_{21}u_1(t) + \cdots + b_{2p}u_p(t)\\
&\vdots\\
\dot{x}_n(t) &= a_{n1}x_1(t) + a_{n2}x_2(t) + \cdots + a_{nn}x_n(t) + b_{n1}u_1(t) + \cdots + b_{np}u_p(t)
\end{aligned}\right\} \tag{1.1}$$

其中 $\dot{x} = \frac{\mathrm{d}x}{\mathrm{d}t}$。这个微分方程组可以写成如下的矩阵形式

$$\begin{bmatrix}\dot{x}_1\\ \dot{x}_2\\ \vdots\\ \dot{x}_n\end{bmatrix} = \begin{bmatrix}a_{11} & a_{12} & \cdots & a_{1n}\\ a_{21} & a_{22} & \cdots & a_{2n}\\ \vdots & \vdots & & \vdots\\ a_{n1} & a_{n2} & \cdots & a_{nn}\end{bmatrix}\begin{bmatrix}x_1\\ x_2\\ \vdots\\ x_n\end{bmatrix} + \begin{bmatrix}b_{11} & b_{12} & \cdots & b_{1p}\\ b_{21} & b_{22} & \cdots & b_{2p}\\ \vdots & \vdots & & \vdots\\ b_{n1} & b_{n2} & \cdots & b_{np}\end{bmatrix}\begin{bmatrix}u_1\\ u_2\\ \vdots\\ u_p\end{bmatrix} \tag{1.2}$$

由状态变量组成的列向量称为状态向量,并记为

$$\boldsymbol{x}=\begin{bmatrix}x_1\\x_2\\\vdots\\x_n\end{bmatrix}$$

输入信号的向量记为 $\boldsymbol{u}$,这样系统就可以用状态微分方程的向量-矩阵形式表示成

$$\dot{\boldsymbol{x}}(t)=\boldsymbol{A}\boldsymbol{x}(t)+\boldsymbol{B}\boldsymbol{u}(t) \tag{1.3}$$

式(1.3)通常也称为状态方程。$n\times n$ 方阵 $\boldsymbol{A}$ 称为系统矩阵,$n\times p$ 矩阵 $\boldsymbol{B}$ 称为输入矩阵。在本书中,矩阵用黑体大写字母表示,而向量则用黑体小写字母表示。

一般情况下,一个线性系统的输出可以通过输出方程

$$\begin{bmatrix}y_1\\y_2\\\vdots\\y_q\end{bmatrix}=\begin{bmatrix}c_{11}&c_{12}&\cdots&c_{1n}\\c_{21}&c_{22}&\cdots&c_{2n}\\\vdots&\vdots&&\vdots\\c_{q1}&c_{q2}&\cdots&c_{qn}\end{bmatrix}\begin{bmatrix}x_1\\x_2\\\vdots\\x_n\end{bmatrix}=\begin{bmatrix}d_{11}&d_{12}&\cdots&d_{1p}\\d_{21}&d_{22}&\cdots&d_{2p}\\\vdots&\vdots&&\vdots\\d_{q1}&d_{q2}&\cdots&d_{qp}\end{bmatrix}\begin{bmatrix}u_1\\u_2\\\vdots\\u_p\end{bmatrix}$$

即

$$\boldsymbol{y}(t)=\boldsymbol{C}\boldsymbol{x}(t)+\boldsymbol{D}\boldsymbol{u}(t) \tag{1.4}$$

与状态变量以及输入信号相联系,其中 $\boldsymbol{y}$ 是一组表示成列向量形式的输出信号,$q\times n$ 矩阵 $\boldsymbol{C}$ 称为输出矩阵,$q\times p$ 矩阵 $\boldsymbol{D}$ 称为前馈矩阵或直接传递矩阵。在许多实例中取 $\boldsymbol{D}=\boldsymbol{0}$。

状态方程和输出方程合在一起称为状态空间表达式或状态空间描述。图 1.3 中所示方框图有助于观察了解许多环节和信号之间的关系。双线表示的是向量而不是单变量,积分方框是 n 个积分器的符号,每个状态变量一个积分器。

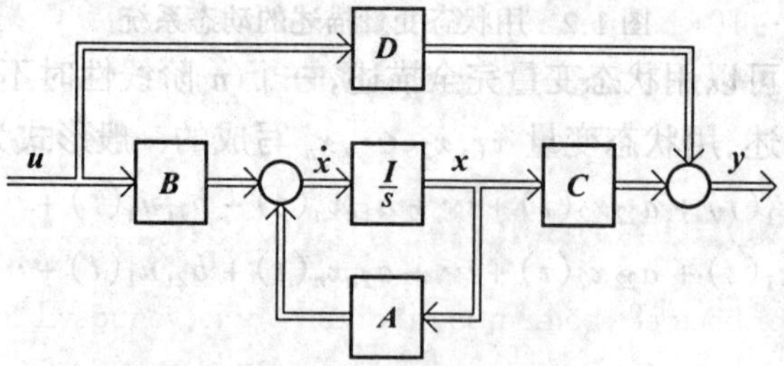

图 1.3　一般状态空间描述的方框图

至于线性时变系统,状态空间表达式的标准形式为

$$\dot{\boldsymbol{x}}(t)=\boldsymbol{A}(t)\boldsymbol{x}(t)+\boldsymbol{B}(t)\boldsymbol{u}(t)$$

$$\boldsymbol{y}(t)=\boldsymbol{C}(t)\boldsymbol{x}(t)+\boldsymbol{D}(t)\boldsymbol{u}(t)$$

其中矩阵 $\boldsymbol{A}(t)$,$\boldsymbol{B}(t)$,$\boldsymbol{C}(t)$和 $\boldsymbol{D}(t)$为时间的连续函数。

1.2　由输入 - 输出表达式建立状态空间表达式

在实际应用中,经常需要由输入 - 输出表达式,即描述系统动态特性的微分方程建立状态空间表达式。对于线性时不变系统,由相应的传递函数可以导出相同的结果。在本节中,将仅考虑线性时不变单变量系统,因此输入和输出都将是标量。由于一个给定系统的状态变量不是唯一的,在一般情况下更愿意建立的是一些“特殊的”,或者说规范形的状态空间表达式。

1.2.1　输入不含导数项

一般而言,在输入不含导数项的情况下,一个 n 阶线性时不变系统可以用以下微分方程描述

$$y^{(n)}(t)+a_1y^{(n-1)}(t)+\cdots+a_{n-1}\dot{y}(t)+a_ny(t)=u(t) \tag{1.5}$$

其中 $y(t)$为输出量,$u(t)$为输入量。相应的传递函数由下式给出为

$$\frac{Y(s)}{U(s)}=\frac{1}{s^n+a_1s^{n-1}+\cdots+a_{n-1}s+a_n} \tag{1.6}$$

1.能控规范形

对于上面提出的情况,定义状态变量如下

$$\left.\begin{aligned} x_1(t)&=y(t)\\ x_2(t)&=\dot{y}(t)\\ &\vdots\\ x_n(t)&=y^{(n-1)}(t)\end{aligned}\right\} \tag{1.7}$$

然后就可以获得一组一阶微分方程

$$\left.\begin{aligned} \dot{x}_1(t)&=x_2(t)\\ \dot{x}_2(t)&=x_3(t)\\ &\vdots\\ \dot{x}_n(t)&=-a_nx_1(t)-a_{n-1}x_2(t)-\cdots-a_1x_n(t)+u(t)\end{aligned}\right\} \tag{1.8}$$

使式(1.5)的最高次导数项等于其余部分可以得到式(1.8)中的最后一个方程。输出方程很简单,就是

$$y(t)=x_1(t) \tag{1.9}$$

用向量 - 矩阵形式表示时,状态空间表达式写成

$$\left.\begin{aligned} \dot{\boldsymbol{x}}(t)&=\left[\begin{array}{c|ccc} 0 & & & \\ \vdots & & \boldsymbol{I}_{n-1} & \\ 0 & & & \\ \hline -a_n & -a_{n-1} & \cdots & -a_1\end{array}\right]\boldsymbol{x}(t)+\begin{bmatrix}0\\ \vdots\\ 0\\ 1\end{bmatrix}u(t)\\ y(t)&=\begin{bmatrix}1 & 0 & \cdots & 0\end{bmatrix}\boldsymbol{x}(t)\end{aligned}\right\} \tag{1.10}$$

由于随后将要讨论的原因,如果一个状态空间表达式具有式(1.10)所示的一对系统

矩阵和输入矩阵,则称为是能控规范形。注意,这时的系统矩阵是一个伴随矩阵,而且在微分方程或传递函数中出现的系数也直接出现在系统矩阵中。实际上很明显,这一能控规范形可以直接由给定的微分方程或传递函数写出。

图 1.4 所示方框图是根据式(1.10)中给出的状态空间表达式画出的,并称为状态变量图。状态变量图由积分器、乘法器、加法器及信号线组成,每一个积分器的输出都是一个状态变量。

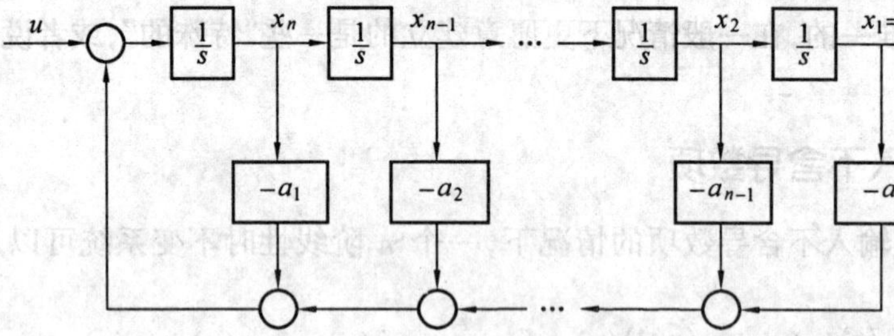

图 1.4　能控规范形的状态变量图

一个具有预先规定的有理传递函数的线性时不变状态空间表达式称为该传递函数的一个实现。在现实世界中基于状态空间表达式可以利用运算放大器电路构造具有这一传递函数的系统,因此用"实现"这个术语来表示。

2. 能观测规范形

在非零初始条件下对式 (1.5) 进行拉普拉斯变换,并将 s 幂次相同的项合并在一起,可以得到

$$\begin{aligned}&(s^n + a_1 s^{n-1} + \cdots + a_n)Y(s) = \\ &U(s) + y(0)s^{n-1} + [\dot{y}(0) + a_1 y(0)]s^{n-2} + \cdots + \\ &[y^{(n-2)}(0) + a_1 y^{(n-3)}(0) + \cdots + a_{n-2}y(0)]s + \\ &[y^{(n-1)}(0) + a_1 y^{(n-2)}(0) + \cdots + a_{n-1}y(0)]\end{aligned}$$

基于上述方程,通过观察可以选择状态变量为

$$\left.\begin{aligned}x_n &= y \\ x_{n-1} &= \dot{y} + a_1 y \\ &\vdots \\ x_2 &= y^{(n-2)} + a_1 y^{(n-3)} + \cdots + a_{n-2} y \\ x_1 &= y^{(n-1)} + a_1 y^{(n-2)} + \cdots + a_{n-1} y\end{aligned}\right\} \tag{1.11}$$

这样,由式(1.5)和式(1.11)得到另一组一阶微分方程

$$\left.\begin{aligned}\dot{x}_1 &= u - a_n y = u - a_n x_n \\ \dot{x}_2 &= x_1 - a_{n-1} y = x_1 - a_{n-1} x_n \\ &\vdots \\ \dot{x}_{n-1} &= x_{n-2} - a_2 x_n \\ \dot{x}_n &= x_{n-1} - a_1 x_n\end{aligned}\right\} \tag{1.12}$$

输出方程为

$$y = x_n \tag{1.13}$$

用矩阵形式表示则状态空间表达式为

$$\left.\begin{aligned}\dot{\boldsymbol{x}} &= \left[\begin{array}{ccc:c} 0 & \cdots & 0 & -a_n \\ \hdashline & & & -a_{n-1} \\ & I_{n-1} & & \vdots \\ & & & -a_1 \end{array}\right]\boldsymbol{x} + \begin{bmatrix} 1 \\ 0 \\ \vdots \\ 0 \end{bmatrix} u \\ y &= [0 \quad \cdots \quad 0 \quad 1]\boldsymbol{x}\end{aligned}\right\} \tag{1.14}$$

由于随后将要讨论的原因，如果一个状态空间表达式具有式(1.14)所示的一对系统矩阵和输出矩阵，则称为是能观测规范形。而且，这一规范形显然可以直接由给定的微分方程或传递函数写出。图 1.5 为能观测规范形状态变量图。

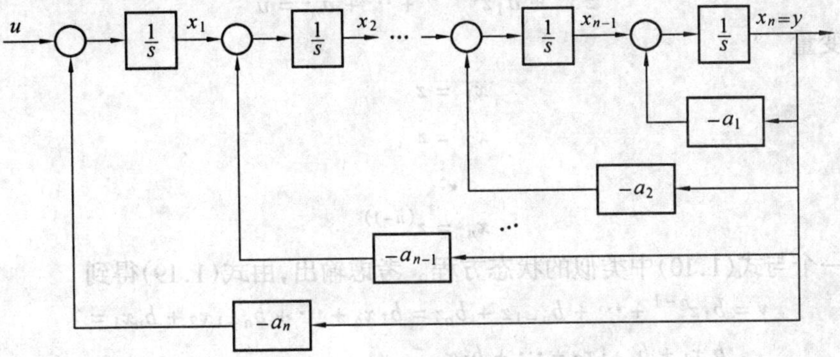

图 1.5　能观测规范形状态变量图

例 1.1　考虑以下微分方程

$$\ddot{y} + 3\dot{y} + 2y = u$$

分别写出它的能控规范形和能观测规范形的状态空间表达式。

解　定义 $x_1 = y$ 和 $x_2 = \dot{y}$，则可得到能控规范形的状态空间表达式

$$\dot{\boldsymbol{x}} = \begin{bmatrix} 0 & 1 \\ -2 & -3 \end{bmatrix}\boldsymbol{x} + \begin{bmatrix} 0 \\ 1 \end{bmatrix} u$$

$$y = [1 \quad 0]\boldsymbol{x}$$

类似地，取 $x_1 = \dot{y} + 3y$ 和 $x_2 = y$ 可以有能观测规范形的状态空间表达式

$$\dot{\boldsymbol{x}} = \begin{bmatrix} 0 & -2 \\ 1 & -3 \end{bmatrix}\boldsymbol{x} + \begin{bmatrix} 1 \\ 0 \end{bmatrix} u$$

$$y = [0 \quad 1]\boldsymbol{x}$$

1.2.2　输入含有导数项

现在考虑由以下微分方程描述的系统

$$y^{(n)} + a_1 y^{(n-1)} + \cdots + a_{n-1}\dot{y} + a_n y = b_1 u^{(n-1)} + \cdots + b_{n-1}\dot{u} + b_n u \tag{1.15}$$

式中 $y(t)$ 为输出量，$u(t)$ 为输入量。该系统相应的传递函数为

$$\frac{Y(s)}{U(s)}=\frac{b_1s^{n-1}+\cdots+b_{n-1}s+b_n}{s^n+a_1s^{n-1}+\cdots+a_{n-1}s+a_n} \tag{1.16}$$

1.能控规范形

将式(1.16)的传递函数改写为

$$\frac{Y(s)}{U(s)}=\frac{Y(s)}{Z(s)}\cdot\frac{Z(s)}{U(s)} \tag{1.17}$$

其中

$$\frac{Z(s)}{U(s)}=\frac{1}{s^n+a_1s^{n-1}+\cdots+a_{n-1}s+a_n} \tag{1.18}$$

$$\frac{Y(s)}{Z(s)}=b_1s^{n-1}+\cdots+b_{n-1}s+b_n \tag{1.19}$$

由式(1.18)有

$$z^{(n)}+a_1z^{(n-1)}+\cdots+a_nz=u \tag{1.20}$$

选择状态变量

$$\begin{aligned} x_1&=z\\ x_2&=\dot{z}\\ &\vdots\\ x_n&=z^{(n-1)} \end{aligned}$$

结果得到一个与式(1.10)中类似的状态方程。考虑输出,由式(1.19)得到

$$\begin{aligned} y=b_1z^{n-1}+\cdots+b_{n-1}\dot{z}+b_nz=b_1x_n+\cdots+b_{n-1}x_2+b_nx_1=\\ b_nx_1+b_{n-1}x_2+\cdots+b_1x_n \end{aligned} \tag{1.21}$$

用矩阵形式表示时,状态空间表达式由下式给出

$$\left.\begin{aligned} \dot{\boldsymbol{x}}&=\left[\begin{array}{c:ccc} 0 & & & \\ \vdots & & I_{n-1} & \\ 0 & & & \\ \hdashline -a_n & -a_{n-1} & \cdots & -a_1 \end{array}\right]\boldsymbol{x}+\begin{bmatrix}0\\ \vdots\\ 0\\ 1\end{bmatrix}u\\ y&=[b_n\quad b_{n-1}\quad\cdots\quad b_1]\boldsymbol{x} \end{aligned}\right\} \tag{1.22}$$

显然,这一状态空间表达式是能控规范形的,而且也可以直接由式(1.15)或式(1.16)写出。

2.能观测规范形

按1.2.1节中类似的方法,在非零初始条件下对式(1.15)取拉普拉斯变换并将 s 幂次相同的项合并在一起可以得到

$$\begin{aligned} &(s^n+a_1s^{n-1}+\cdots+a_n)Y(s)=\\ &(b_1s^{n-1}+b_2s^{n-2}+\cdots+b_n)U(s)+y(0)s^{n-1}+[\dot{y}(0)+a_1y(0)-b_1u(0)]s^{n-2}+\cdots+\\ &[y^{(n-2)}(0)+a_1y^{(n-3)}(0)+\cdots+a_{n-2}y(0)-b_1u^{(n-3)}(0)-\cdots-b_{n-2}u(0)]s+\\ &[y^{(n-1)}(0)+a_1y^{(n-2)}(0)+\cdots+a_{n-1}y(0)-b_1u^{(n-2)}(0)-\cdots-b_{n-1}u(0)] \end{aligned} \tag{1.23}$$

选择状态变量

$$\left.\begin{aligned} x_n &= y \\ x_{n-1} &= \dot{y} + a_1 y - b_1 u \\ &\vdots \\ x_2 &= y^{(n-2)} + a_1 y^{(n-3)} + \cdots + a_{n-2} y - b_1 u^{(n-3)} - \cdots - b_{n-2} u \\ x_1 &= y^{(n-1)} + a_1 y^{(n-2)} + \cdots + a_{n-1} y - b_1 u^{(n-2)} - \cdots - b_{n-1} u \end{aligned}\right\} \tag{1.24}$$

结果得到能观测规范形的状态空间表达式

$$\left.\begin{aligned} \dot{\boldsymbol{x}} &= \left[\begin{array}{ccc:c} 0 & \cdots & 0 & -a_n \\ \hdashline & & & -a_{n-1} \\ & \boldsymbol{I}_{n-1} & & \vdots \\ & & & -a_1 \end{array}\right] \boldsymbol{x} + \begin{bmatrix} b_n \\ b_{n-1} \\ \vdots \\ b_1 \end{bmatrix} u \\ y &= [0 \ \cdots \ 0 \ 1] \boldsymbol{x} \end{aligned}\right\} \tag{1.25}$$

例 1.2　考虑传递函数

$$\frac{Y(s)}{U(s)} = \frac{11s + 6}{s^3 + 6s^2 + 11s + 6}$$

分别写出能控规范形和能观测规范形的状态空间表达式。

解　令 $\dddot{z} + 6\ddot{z} + 11\dot{z} + 6z = u$，并选择状态变量为

$$x_1 = z$$
$$x_2 = \dot{z}$$
$$x_3 = \ddot{z}$$

则得到能控规范形的状态空间表达式

$$\dot{\boldsymbol{x}} = \begin{bmatrix} 0 & 1 & 0 \\ 0 & 0 & 1 \\ -6 & -11 & -6 \end{bmatrix} \boldsymbol{x} + \begin{bmatrix} 0 \\ 0 \\ 1 \end{bmatrix} u$$

$$y = [6 \ \ 11 \ \ 0] \boldsymbol{x}$$

选择状态变量为

$$x_3 = y$$
$$x_2 = \dot{y} + 6y$$
$$x_1 = \ddot{y} + 6\dot{y} + 11y - 11u$$

则获得能观测规范形的状态空间表达式

$$\dot{\boldsymbol{x}} = \begin{bmatrix} 0 & 0 & -6 \\ 1 & 0 & -11 \\ 0 & 1 & -6 \end{bmatrix} \boldsymbol{x} + \begin{bmatrix} 6 \\ 11 \\ 0 \end{bmatrix} u$$

$$y = [0 \ \ 0 \ \ 1] \boldsymbol{x}$$

3. 约当规范形

如果可以把状态方程的系统矩阵写成约当形，有时候会特别有用。相应地，这种状态

空间表达式是约当规范形。

情况 1 传递函数的所有极点互异

首先考虑最简单的情况，由式(1.16) 给出的传递函数的所有极点 $s_1, s_2, \cdots, s_n$ 互异。这时该传递函数可以通过部分分式分解展开为

$$\frac{Y(s)}{U(s)} = \frac{M(s)}{(s - s_1)(s - s_2)\cdots(s - s_n)} = \sum_{i=1}^{n} \frac{c_i}{s - s_i}$$

其中

$$c_i = \lim_{s \to s_i}(s - s_i)\frac{Y(s)}{U(s)}$$

选择状态变量获得典型函数

$$X_i(s) = \frac{1}{s - s_i}U(s), \qquad i = 1,2,\cdots,n \tag{1.26}$$

由此

$$sX_i(s) = s_i X_i(s) + U(s)$$

或

$$\dot{x}(t) = s_i x_i(t) + u(t)$$

输出则为

$$y(t) = \sum_{i=1}^{n} c_i x_i(t)$$

这样，状态空间表达式就写成

$$\left.\begin{aligned} \dot{\boldsymbol{x}} &= \begin{bmatrix} s_1 & 0 & \cdots & 0 \\ 0 & s_2 & \cdots & 0 \\ \vdots & \vdots & & \vdots \\ 0 & 0 & \cdots & s_n \end{bmatrix}\boldsymbol{x} + \begin{bmatrix} 1 \\ 1 \\ \vdots \\ 1 \end{bmatrix}u \\ y &= [c_1 \quad c_2 \quad \cdots \quad c_n]\boldsymbol{x} \end{aligned}\right\} \tag{1.27}$$

注意，在系统矩阵中主对角线元素正好是该系统的极点，而且只有主对角线元素才有可能是非零的。在这种情况下，由于所有极点互异，系统矩阵变为一个对角阵($\boldsymbol{\Lambda}$ 阵)。

图 1.6 为式(1.27) 的状态变量图。

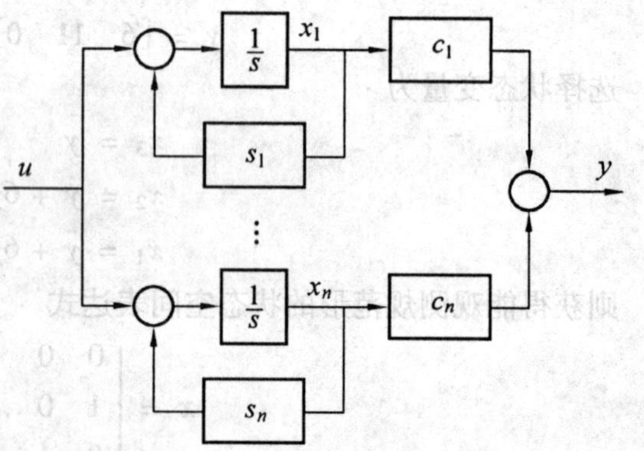

图 1.6 式(1.27) 的状态变量图

例 1.3 仍然考虑例 1.2 中给出的传递函数

$$\frac{Y(s)}{U(s)} = \frac{11s + 6}{s^3 + 6s^2 + 11s + 6}$$

写出它的约当规范形状态空间表达式。

解 传递函数可以通过部分分式分解展开成

$$\frac{Y(s)}{U(s)} = \frac{11s+6}{(s+1)(s+2)(s+3)} = -\frac{5}{2}\frac{1}{s+1} + 16\frac{1}{s+2} - \frac{17}{2}\frac{1}{s+3}$$

约当规范形状态空间表达式为

$$\dot{x} = \begin{bmatrix} -1 & 0 & 0 \\ 0 & -2 & 0 \\ 0 & 0 & -3 \end{bmatrix} x + \begin{bmatrix} 1 \\ 1 \\ 1 \end{bmatrix} u$$

$$y = \begin{bmatrix} -\frac{5}{2} & 16 & -\frac{17}{2} \end{bmatrix} x$$

其中状态变量满足

$$X_1(s) = \frac{1}{s+1}U(s), \quad X_2(s) = \frac{1}{s+2}U(s), \quad X_3(s) = \frac{1}{s+3}U(s)$$

情况 2　传递函数的极点非互异

与前一种情况不一样，如果一个传递函数有重极点，则并不总是能建立具有对角形系统矩阵的状态空间表达式。现将用一些例子说明导出约当规范形的过程。这种思路可以很容易地推广到一般情况。

为了方便起见，假设传递函数的极点中 s_1 为 r 重，而所有其他极点均为互异。这样，传递函数就可以通过部分分式分解展开成

$$\frac{Y(s)}{U(s)} = \sum_{i=1}^{r} \frac{c_{1i}}{(s-s_1)^i} + \sum_{j=r+1}^{n} \frac{c_j}{s-s_j}$$

式中

$$c_{1i} = \frac{1}{(r-i)!} \lim_{s \to s_1} \frac{\mathrm{d}^{r-i}}{\mathrm{d}s^{r-i}} \left[(s-s_1)^r \frac{Y(s)}{U(s)} \right], \qquad i = 1,2,\cdots,r$$

$$c_j = \lim_{s \to s_j} (s-s_j) \frac{Y(s)}{U(s)}, \qquad j = r+1, r+2, \cdots, n$$

选择典型函数为

$$X_i(s) = \frac{1}{(s-s_1)^{r-i+1}} U(s), \qquad i = 1,2,\cdots,r \tag{1.28}$$

$$X_j(s) = \frac{1}{s-s_j} U(s), \qquad j = r+1, r+2, \cdots, n \tag{1.29}$$

然后考虑 $\dot{x}_k(t)$ 的表达式，$k = 1,2,\cdots,r-1,r,r+1,\cdots,n$

(1) $k < r$

$$X_k(s) = \frac{1}{(s-s_1)^{r-k+1}} U(s) = \frac{1}{(s-s_1)(s-s_1)^{r-(k+1)+1}} U(s) = \frac{1}{s-s_1} X_{k+1}(s)$$

$$sX_k(s) = s_1 X_k(s) + X_{k+1}(s)$$

$$\dot{x}_k = s_1 x_k + x_{k+1}$$

(2) $k = r$

$$X_k(s) = \frac{1}{s-s_1} U(s)$$

$$\dot{x}_k = s_1 x_k + u$$

(3) $k > r$

$$X_k(s) = \frac{1}{s - s_k}U(s)$$

$$\dot{x}_k = s_k x_k + u$$

将以上一阶微分方程写成矩阵形式得到状态方程

$$\begin{bmatrix} \dot{x}_1 \\ \dot{x}_2 \\ \vdots \\ \dot{x}_{r-1} \\ \dot{x}_r \\ \dot{x}_{r+1} \\ \vdots \\ \dot{x}_n \end{bmatrix} = \left[\begin{array}{ccccc:ccc} s_1 & 1 & \cdots & 0 & 0 & 0 & \cdots & 0 \\ 0 & s_1 & & 0 & 0 & 0 & \cdots & 0 \\ \vdots & & & & \vdots & \vdots & & \vdots \\ 0 & 0 & \cdots & s_1 & 1 & 0 & \cdots & 0 \\ 0 & 0 & \cdots & 0 & s_1 & 0 & \cdots & 0 \\ \hdashline 0 & 0 & \cdots & 0 & 0 & s_{r+1} & & \\ \vdots & & & & \vdots & & & \\ 0 & 0 & \cdots & 0 & 0 & 0 & \cdots & s_n \end{array}\right] \begin{bmatrix} x_1 \\ x_2 \\ \vdots \\ x_{r-1} \\ x_r \\ x_{r+1} \\ \vdots \\ x_n \end{bmatrix} + \begin{bmatrix} 0 \\ 0 \\ \vdots \\ 0 \\ 1 \\ 1 \\ \vdots \\ 1 \end{bmatrix} u \tag{1.30}$$

至于输出则有

$$Y(s) = \sum_{i=1}^{r} \frac{c_{1i}}{(s - s_1)^i}U(s) + \sum_{j=r+1}^{n} \frac{c_j}{s - s_j}U(s) =$$

$$c_{11}X_r(s) + c_{12}X_{r-1}(s) + \cdots + c_{1r}X_1(s) + c_{r+1}X_{r+1}(s) + \cdots + c_nX_n(s)$$

$$y(t) = c_{1r}x_1(t) + \cdots + c_{11}x_r(t) + c_{r+1}x_{r+1}(t) + \cdots + c_nx_n(t)$$

即矩阵形式的输出方程为

$$y = [\, c_{1r} \quad c_{1,r-1} \cdots \quad c_{12} \quad c_{11} \;\vdots\; c_{r+1} \quad \cdots \quad c_n \,]\boldsymbol{x} \tag{1.31}$$

注意,系统矩阵的主对角线由一些约当块组成而其余元素为零。对角线系统矩阵是约当形的一种特殊情况:系统矩阵所有的约当块均为一阶。图 1.7 为这种情况下的状态变量图。

由于主对角线元素就是系统的极点,也即系统矩阵的特征值,因此约当规范形有时也称为特征值规范形。

例 1.4 确定约当规范形的状态空间表达式,描述系统的传递函数为

$$\frac{Y(s)}{U(s)} = \frac{2s^2 + 6s + 5}{s^3 + 4s^2 + 5s + 2}$$

解 传递函数可以通过部分分式分解展开成

$$\frac{Y(s)}{U(s)} = \frac{2s^2 + 6s + 5}{(s + 1)^2(s + 2)} = \frac{c_{11}}{(s + 1)^2} + \frac{c_{12}}{s + 1} + \frac{c_3}{s + 2}$$

其中

$$c_{12} = \lim_{s \to -1} \frac{2s^2 + 6s + 5}{s + 2} = 1, \; c_{11} = \lim_{s \to -1} \frac{\mathrm{d}}{\mathrm{d}s} \frac{2s^2 + 6s + 5}{s + 2} = 1$$

$$c_3 = \lim_{s \to -2} \frac{2s^2 + 6s + 5}{(s + 1)^2} = 1$$

即

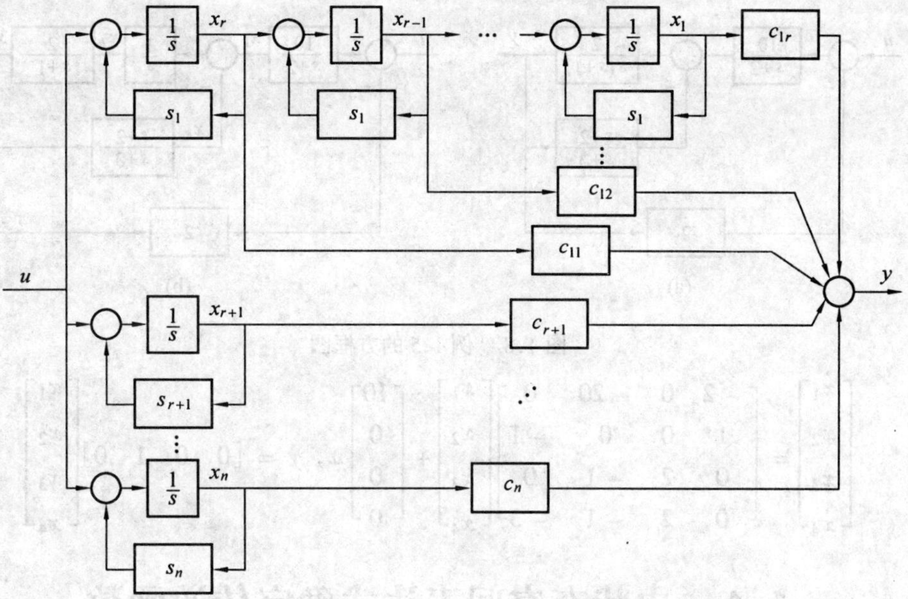

图 1.7　式(1.30) 和式(1.31) 的动态变量图

$$\frac{Y(s)}{U(s)} = \frac{1}{(s+1)^2} + \frac{1}{s+1} + \frac{1}{s+2}$$

由式 (1.30) 有

$$\dot{x} = \left[\begin{array}{cc:c} -1 & 1 & 0 \\ 0 & -1 & 0 \\ \hdashline 0 & 0 & -2 \end{array}\right] x + \left[\begin{array}{c} 0 \\ 1 \\ \hdashline 1 \end{array}\right] u$$

$$y = [1 \quad 1 \;\vdots\; 1] x$$

其中状态变量满足

$$X_1(s) = \frac{1}{(s+1)^2} U(s),\ X_2(s) = \frac{1}{s+1} U(s),\ X_3(s) = \frac{1}{s+2} U(s)$$

1.3　由方框图建立状态空间表达式

有时候直接由给定的方框图建立状态空间表达式,这样就可以使得方框的输出选做状态变量以反映实际的物理量。下面用一个例子说明这一方法。

例 1.5　写出图 1.8(a) 所示系统的状态空间表达式。

解　首先将方框图改画成如图 1.8(b) 所示,使得每一个方框都可以用一个积分器和/或乘法器仿真。

如图 1.8(b) 所示,将含有一阶环节的方框输出选做状态变量,可以得到

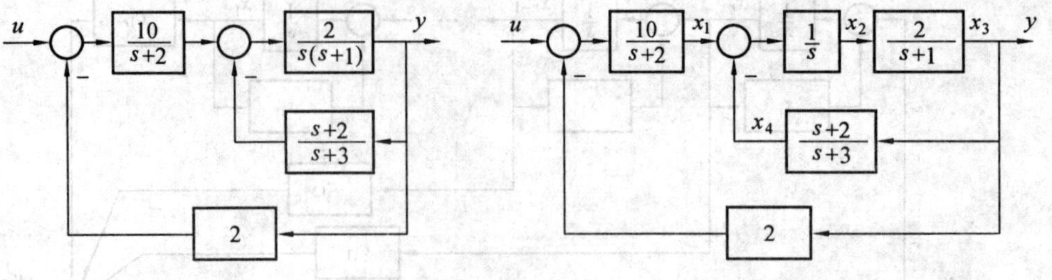

(a) (b)

图 1.8 例 1.5 的方框图

$$\begin{bmatrix}\dot{x}_1\\ \dot{x}_2\\ \dot{x}_3\\ \dot{x}_4\end{bmatrix}=\begin{bmatrix}-2 & 0 & -20 & 0\\ 1 & 0 & 0 & -1\\ 0 & 2 & -1 & 0\\ 0 & 2 & -1 & -3\end{bmatrix}\begin{bmatrix}x_1\\ x_2\\ x_3\\ x_4\end{bmatrix}+\begin{bmatrix}10\\ 0\\ 0\\ 0\end{bmatrix}u,\ y=\begin{bmatrix}0 & 0 & 1 & 0\end{bmatrix}\begin{bmatrix}x_1\\ x_2\\ x_3\\ x_4\end{bmatrix}$$

1.4 由状态空间表达式确定传递函数

当一个系统的微分方程或传递函数已知时,上面给出了写出状态空间表达式的方法。在有些情况下需要由状态空间表达式估算传递函数(对于多变量系统则为传递函数矩阵)。这一问题可以陈述如下:给定系数矩阵 $\boldsymbol{A}$、$\boldsymbol{B}$、$\boldsymbol{C}$ 和 $\boldsymbol{D}$,确定输出 y 与输入 u 之间的关系。

从单变量系统状态空间表达式的典型形式着手

$$\left.\begin{aligned}\dot{\boldsymbol{x}}(t)&=\boldsymbol{Ax}(t)+\boldsymbol{B}u(t)\\ y(t)&=\boldsymbol{Cx}(t)\end{aligned}\right\}\tag{1.32}$$

令初始条件为零,对式(1.32)取拉普拉斯变换得到

$$sX(s)=\boldsymbol{A}X(s)+\boldsymbol{B}U(s)\tag{1.33}$$

这一方程可以写成

$$(s\boldsymbol{I}-\boldsymbol{A})X(s)=\boldsymbol{B}U(s)$$

并很容易求解出

$$X(s)=(s\boldsymbol{I}-\boldsymbol{A})^{-1}\boldsymbol{B}U(s)\tag{1.34}$$

将以上 $X(s)$ 的表达式代入式(1.32)中的输出方程得到

$$Y(s)=\boldsymbol{C}(s\boldsymbol{I}-\boldsymbol{A})^{-1}\boldsymbol{B}U(s)\tag{1.35}$$

因此,系统的传递函数由下式给出

$$G(s)=\frac{Y(s)}{U(s)}=\boldsymbol{C}(s\boldsymbol{I}-\boldsymbol{A})^{-1}\boldsymbol{B}\tag{1.36}$$

例 1.6 确定由以下状态空间表达式描述的系统的传递函数

$$\dot{\boldsymbol{x}}=\begin{bmatrix}0 & 1 & 0\\ 0 & -1 & -1\\ 0 & 0 & -3\end{bmatrix}\boldsymbol{x}+\begin{bmatrix}0\\ 1\\ 1\end{bmatrix}u$$

$$y=\begin{bmatrix}1 & 0 & 0\end{bmatrix}\boldsymbol{x}$$

解　由于

$$\det(s\boldsymbol{I}-\boldsymbol{A})=\begin{vmatrix} s & -1 & 0 \\ 0 & s+1 & 1 \\ 0 & 0 & s+3 \end{vmatrix}=s(s+1)(s+3)$$

$$\mathrm{adj}(s\boldsymbol{I}-\boldsymbol{A})=\begin{bmatrix} (s+1)(s+3) & s+3 & -1 \\ 0 & s(s+3) & -s \\ 0 & 0 & s(s+1) \end{bmatrix}$$

有

$$G(s)=\frac{Y(s)}{U(s)}=\boldsymbol{C}(s\boldsymbol{I}-\boldsymbol{A})^{-1}\boldsymbol{B}=\boldsymbol{C}\frac{\mathrm{adj}(s\boldsymbol{I}-\boldsymbol{A})}{\det(s\boldsymbol{I}-\boldsymbol{A})}\boldsymbol{B}=$$

$$[1\ \ 0\ \ 0]\frac{1}{s(s+1)(s+3)}\begin{bmatrix} (s+1)(s+3) & s+3 & -1 \\ 0 & s(s+3) & -s \\ 0 & 0 & s(s+1) \end{bmatrix}\begin{bmatrix} 0 \\ 1 \\ 1 \end{bmatrix}=$$

$$\frac{s+2}{s(s+1)(s+3)}$$

对于多变量的情况，输入 $\boldsymbol{U}(s)$ 是一个 $p\times 1$ 向量，输出 $\boldsymbol{Y}(s)$ 是一个 $q\times 1$ 向量。前面的推导继续有效，只是式(1.35) 变为

$$\boldsymbol{Y}(s)=\boldsymbol{C}(s\boldsymbol{I}-\boldsymbol{A})^{-1}\boldsymbol{B}\boldsymbol{U}(s)=\boldsymbol{G}(s)\boldsymbol{U}(s)$$

即

$$\boldsymbol{G}(s)=\boldsymbol{C}(s\boldsymbol{I}-\boldsymbol{A})^{-1}\boldsymbol{B} \tag{1.37}$$

其中 $\boldsymbol{G}(s)$ 现在是一个 $q\times p$ 矩阵。$\boldsymbol{G}(s)$ 的元素 $G_{ij}(s)$ 是其他所有输入均为零时得到的传递函数

$$G_{ij}(s)=\frac{Y_i(s)}{U_j(s)} \tag{1.38}$$

1.5　特征多项式和特征值

对于一个特定的系统，状态变量的选择不是唯一的。因此一个系统的状态空间表达式也不是唯一的。

先考虑线性非奇异变换对状态空间表达式的影响。假设某系统由以下的状态空间表达式描述

$$\left.\begin{aligned} \dot{\boldsymbol{x}}(t)&=\boldsymbol{A}\boldsymbol{x}(t)+\boldsymbol{B}\boldsymbol{u}(t) \\ \boldsymbol{y}(t)&=\boldsymbol{C}\boldsymbol{x}(t) \end{aligned}\right\} \tag{1.39}$$

可以由原来的状态变量 $\boldsymbol{x}$ 构造一组新的状态变量 $\boldsymbol{z}$

$$\boldsymbol{x}=\boldsymbol{P}\boldsymbol{z} \tag{1.40}$$

变换式中 $\boldsymbol{P}$ 是一个 $n\times n$ 的非奇异矩阵。以 $\boldsymbol{z}$ 代入状态方程得到

$$\boldsymbol{P}\dot{\boldsymbol{z}}=\boldsymbol{A}\boldsymbol{P}\boldsymbol{z}+\boldsymbol{B}\boldsymbol{u}$$

或

$$\dot{z} = P^{-1}APz + P^{-1}Bu \tag{1.41}$$

类似地，以 z 代入输出方程得到

$$y = CPz \tag{1.42}$$

由此，我们可以用新的状态变量写出状态方程和输出方程为

$$\begin{aligned} \dot{z} &= \bar{A}z + \bar{B}u \\ y &= \bar{C}z \end{aligned} \tag{1.43}$$

其中

$$\bar{A} = P^{-1}AP, \quad \bar{B} = P^{-1}B, \quad \bar{C} = CP$$

对于给定的状态空间表达式，可以将式(1.37) 写成

$$G(s) = C\frac{\mathrm{adj}(sI - A)}{\det(sI - A)}B \tag{1.44}$$

其中 $G(s)$ 的分母 $\det(sI - A)$ 称为该系统矩阵的特征多项式。

下面将说明非奇异线性变换不会改变系统矩阵的特征多项式。对于式(1.43) 中新的系统矩阵的特征多项式有

$$\begin{aligned} \det(sI - \bar{A}) &= \det(sI - P^{-1}AP) = \det(sP^{-1}P - P^{-1}AP) = \\ &\det[P^{-1}(sI - A)P] = \det(P^{-1}P) \cdot \det(sI - A) = \\ &\det(sI - A) \end{aligned}$$

这一结果证实系统矩阵的特征多项式与所选择用以表示该系统的状态变量无关。事实上，由式(1.44) 可知，系统矩阵的特征多项式就是 $G(s)$ 的特征多项式。

在状态空间描述中，特征多项式的根，即传递函数的极点称为系统矩阵的特征根或系统的特征根。由此，可以得到一个重要的结论，系统的特征根与所选择的状态变量无关。

习　题

题 1.1　描述电枢控制直流电动机用以下微分方程

$$v_a = L_a\frac{di_a}{dt} + R_a i_a + e_b$$

$$e_b = C_e\frac{d\theta_m}{dt}$$

$$T_m = C_m i_a$$

$$T_m = J_m\frac{d^2\theta_m}{dt^2} + f_m\frac{d\theta_m}{dt}$$

或传递函数

$$\frac{\theta_m(s)}{V_a(s)} = \frac{C_m}{s[L_aJ_ms^2 + (L_af_m + R_aJ_m)s + (R_af_m + C_eC_m)]}$$

(a) 取状态变量 $x_1 = \theta_m, x_2 = \dot{\theta}_m, x_3 = \ddot{\theta}_m$，输出 $y = \theta_m$。写出矩阵形式的状态方程。

(b) 取状态变量 $\bar{x}_1 = i_a, \bar{x}_2 = \theta_m, \bar{x}_3 = \dot{\theta}_m$，输出 $y = \theta_m$。写出矩阵形式的状态方程。

(c) 确定使 $x = T\bar{x}$ 变换矩阵。

题 1.2　如题 1.2 图所示，电路有两个输入：一个电压源和一个电流源。取状态变量 $x_1 = i_L$、$x_2 = v_{C_1}$、$x_3 = v_{C_2}$ 和输出 $y = \theta_m$。确定所示电路的状态空间表达式。

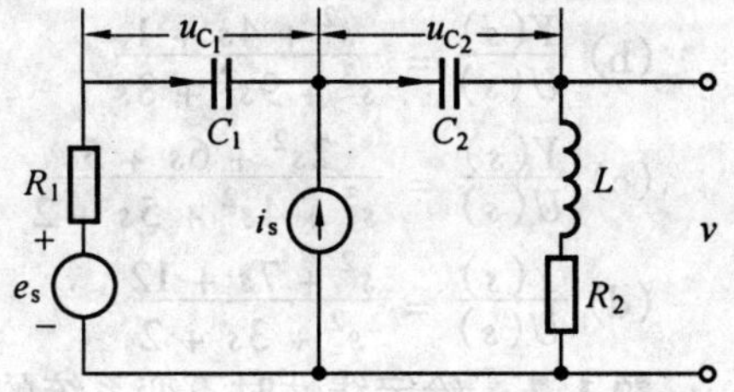

题 1.2 图

题 1.3　系统的方框图如题 1.3 图所示，其中 x_1、x_2 和 x_3 是指定的状态变量。确定各系统的状态空间表达式。

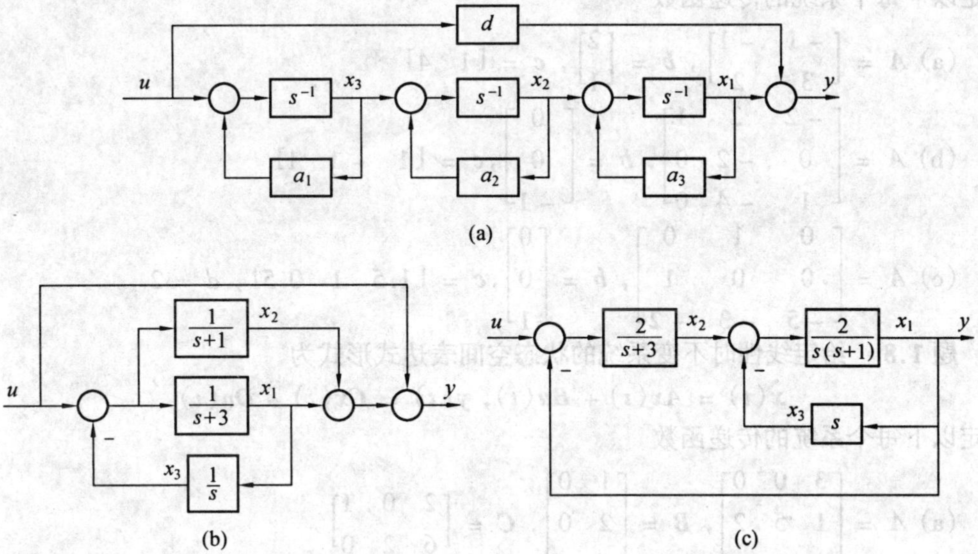

题 1.3 图

题 1.4　求以下系统的状态空间表达式

(a) $\dddot{y} + 3\ddot{y} + 2\dot{y} + 2y = u$

(b) $\dddot{y} - 3\ddot{y} + 4\dot{y} - 2y = 2u$

(c) $\dddot{y} + 3\ddot{y} + 2\dot{y} + 2y = 0$

(d) $\dddot{y} - 3a\ddot{y} + 3a^2\dot{y} - a^3y = u$

(e) $\begin{cases} \ddot{y}_1 + 4\dot{y}_1 - 3y_2 = u_1 \\ \dot{y}_2 + \dot{y}_1 + y_1 + 2y_2 = u_2 \end{cases}$

题 1.5　求以下系统的状态空间表达式

(a) $2\dddot{y} - 3y = \ddot{u} - 2u$

(b) $\dddot{y} + 4\ddot{y} + 2\dot{y} + y = \ddot{u} + \dot{u} + 3u$

(c) $\dddot{y} + 10\ddot{y} + 80\dot{y} + 128y = 5\dot{u} + 6u$

(d) $\dddot{y} + 6\ddot{y} + 11\dot{y} + 6y = \dddot{u} + 8\ddot{u} + 17\dot{u} + u$

题 1.6　确定由以下传递函数所描述系统的状态空间表达式

(a) $\dfrac{Y(s)}{U(s)} = \dfrac{2s^2 + 18s + 40}{s^3 + 6s^2 + 11s + 6}$

(b) $\dfrac{Y(s)}{U(s)} = \dfrac{s^2 + 4s + 1}{s^3 + 9s^2 + 8s}$

(c) $\dfrac{Y(s)}{U(s)} = \dfrac{2s^2 + 6s + 5}{s^3 + 4s^2 + 5s + 2}$

(d) $\dfrac{Y(s)}{U(s)} = \dfrac{s^2 + 7s + 12}{s^2 + 3s + 2}$

题 1.7　给定线性时不变系统的状态空间表达式形式为

$$\dot{x}(t) = Ax(t) + bu(t),\quad y(t) = cx(t) + du(t)$$

确定以下每个系统的传递函数

(a) $A = \begin{bmatrix} -1 & -1 \\ 3 & 2 \end{bmatrix},\ b = \begin{bmatrix} 2 \\ 1 \end{bmatrix},\ c = [1 \quad 4]$

(b) $A = \begin{bmatrix} -2 & 2 & 1 \\ 0 & -2 & 0 \\ 1 & -4 & 0 \end{bmatrix},\ b = \begin{bmatrix} 0 \\ 0 \\ -1 \end{bmatrix},\ c = [1 \quad -1 \quad 1]$

(c) $A = \begin{bmatrix} 0 & 1 & 0 \\ 0 & 0 & 1 \\ -5 & -3 & -2 \end{bmatrix},\ b = \begin{bmatrix} 0 \\ 0 \\ 1 \end{bmatrix},\ c = [1.5 \quad 1 \quad 0.5],\ d = 2$

题 1.8　给定线性时不变系统的状态空间表达式形式为

$$\dot{x}(t) = Ax(t) + Bu(t),\ y(t) = Cx(t) + Du(t)$$

确定以下每个系统的传递函数

(a) $A = \begin{bmatrix} 3 & 0 & 0 \\ 1 & 5 & 2 \\ 0 & 2 & 1 \end{bmatrix},\ B = \begin{bmatrix} 1 & 0 \\ 2 & 0 \\ 0 & 5 \end{bmatrix},\ C = \begin{bmatrix} 2 & 0 & 1 \\ 6 & 2 & 0 \end{bmatrix}$

(b) $A = \begin{bmatrix} 0 & 1 & 0 \\ 0 & 0 & 1 \\ -1 & -1 & -2 \end{bmatrix},\ B = \begin{bmatrix} 0 & 0 \\ 1 & 0 \\ 0 & 1 \end{bmatrix},\ C = \begin{bmatrix} 1 & 0 & 0 \\ 0 & 0 & 1 \end{bmatrix},\ D = \begin{bmatrix} 0 & 1 \\ 1 & 1 \end{bmatrix}$

第 2 章　线性系统的运动分析

在前一章中介绍了控制系统的状态空间描述。显然，在系统分析中下一步是寻找状态方程的解。

2.1　时不变连续系统

2.1.1　齐次方程的解

首先考虑线性时不变连续系统状态方程的解。由于一个微分方程齐次部分的解基本上也确定了该微分方程非齐次部分的解，因此从以下的齐次方程开始

$$\dot{\boldsymbol{x}}(t) = \boldsymbol{A}\boldsymbol{x}(t),\ t \geqslant 0;\ \boldsymbol{x}(0) = \boldsymbol{x}_0 \tag{2.1}$$

和通常在时不变系统的情况下一样，这里假设初始时间为 $t_0 = 0$。

1.无穷级数解

求解微分方程的一种方法是假设方程的解是个系数待定的无穷级数，然后将该无穷级数代入微分方程估算待定系数。下面就用这一方法寻找状态方程的解。

采用这种方法时，方程解的形式必定可以假设为

$$\boldsymbol{x}(t) = \boldsymbol{b}_0 + \boldsymbol{b}_1 t + \boldsymbol{b}_2 t^2 + \cdots + \boldsymbol{b}_k t^k + \cdots = \sum_{k=0}^{\infty} \boldsymbol{b}_k t^k,\ t \geqslant 0 \tag{2.2}$$

将式(2.2) 代入齐次方程得到

$$\boldsymbol{b}_1 + 2\boldsymbol{b}_2 t + 3\boldsymbol{b}_3 t^2 + \cdots + k\boldsymbol{b}_k t^{k-1} + \cdots = \boldsymbol{A}(\boldsymbol{b}_0 + \boldsymbol{b}_1 t + \boldsymbol{b}_2 t^2 + \cdots + \boldsymbol{b}_k t^k + \cdots) \tag{2.3}$$

使式(2.3) 中各 t^i 项的系数相等，$i = 1,2,\cdots$，结果得到

$$\boldsymbol{b}_1 = \boldsymbol{A}\boldsymbol{b}_0,\ \boldsymbol{b}_2 = \frac{1}{2}\boldsymbol{A}\boldsymbol{b}_1 = \frac{1}{2}\boldsymbol{A}^2\boldsymbol{b}_0,\ \boldsymbol{b}_3 = \frac{1}{3}\boldsymbol{A}\boldsymbol{b}_2 = \frac{1}{3!}\boldsymbol{A}^3\boldsymbol{b}_0,\cdots,\boldsymbol{b}_k = \frac{1}{k!}\boldsymbol{A}^k\boldsymbol{b}_0,\cdots$$

因此，齐次解可以表示为

$$\boldsymbol{x}(t) = \boldsymbol{b}_0 + \boldsymbol{A}\boldsymbol{b}_0 t + \frac{1}{2!}\boldsymbol{A}^2\boldsymbol{b}_0 t^2 + \cdots + \frac{1}{k!}\boldsymbol{A}^k\boldsymbol{b}_0 t^k + \cdots =$$

$$\left(\boldsymbol{I}_n + \boldsymbol{A}t + \frac{1}{2!}\boldsymbol{A}^2 t^2 + \cdots + \frac{1}{k!}\boldsymbol{A}^k t^k + \cdots\right)\boldsymbol{b}_0 \tag{2.4}$$

由式(2.4) 和给定的初始条件得到

$$\boldsymbol{b}_0 = \boldsymbol{x}(t)\big|_{t=0} = \boldsymbol{x}_0$$

所以，齐次解由下式给出

$$\boldsymbol{x}(t) = \left(\boldsymbol{I}_n + \boldsymbol{A}t + \frac{1}{2!}\boldsymbol{A}^2 t^2 + \cdots + \frac{1}{k!}\boldsymbol{A}^k t^k + \cdots\right)\boldsymbol{x}_0 \tag{2.5}$$

2.状态转移矩阵和矩阵指数函数

注意到式(2.5)中的每一项都是一个 $n\times n$ 的矩阵,因此可以把齐次解表示成

$$\dot{\boldsymbol{x}}(t)=\boldsymbol{\Phi}(t)\boldsymbol{x}_0 \tag{2.6}$$

其中 $n\times n$ 矩阵 $\boldsymbol{\Phi}(t)$ 称为状态转移矩阵,因为它完全地规定了从初始时间 $t=0$ 到任一时间 $t\geqslant t_0$ 状态的转移。

不仅如此,由式(2.5)中的无穷级数与标量指数的泰勒级数展开之间的相似性,定义矩阵指数函数为

$$\mathrm{e}^{\boldsymbol{A}t}=\boldsymbol{I}_n+\boldsymbol{A}t+\frac{1}{2!}\boldsymbol{A}^2t^2+\cdots+\frac{1}{k!}\boldsymbol{A}^kt^k+\cdots=\sum_{k=0}^{\infty}\frac{1}{k!}\boldsymbol{A}^kt^k \tag{2.7}$$

或者简称为矩阵指数。这样线性时不变状态方程的齐次解就可以表示成

$$\boldsymbol{x}(t)=\mathrm{e}^{\boldsymbol{A}t}\boldsymbol{x}_0 \tag{2.8}$$

且在这种情况下有

$$\boldsymbol{\Phi}(t)=\mathrm{e}^{\boldsymbol{A}t} \tag{2.9}$$

应当指出,式(2.9)中给出的结论仅适用于线性时不变系统。对于时变系统,可以采用状态转移矩阵使 $\boldsymbol{x}(t)$ 与其初始状态 $\boldsymbol{x}_0$ 相联系,但是矩阵指数函数却仅对时不变系统有效。

3.状态转移矩阵的运算性质

状态转移矩阵 $\boldsymbol{\Phi}(t)$ 或者矩阵指数函数 $\mathrm{e}^{\boldsymbol{A}t}$ 是求解线性时不变状态方程的核心问题。由于 $\boldsymbol{\Phi}(t)$ 或 $\mathrm{e}^{\boldsymbol{A}t}$ 包含描述系统动态特性的完整信息,熟悉以下一些基本运算性质是很有用的。

(1) $\boldsymbol{\Phi}(0)=\boldsymbol{I}$ (2.10)

证明 $\boldsymbol{\Phi}(0)=\mathrm{e}^{\boldsymbol{A}t}|_{t=0}=\boldsymbol{I}$

(2) $\dot{\boldsymbol{\Phi}}(t)=\boldsymbol{A}\boldsymbol{\Phi}(t)=\boldsymbol{\Phi}(t)\boldsymbol{A}$ (2.11)

证明

$$\dot{\boldsymbol{\Phi}}(t)=\frac{\mathrm{d}}{\mathrm{d}t}\mathrm{e}^{\boldsymbol{A}t}=\boldsymbol{A}+\boldsymbol{A}^2t+\frac{1}{2!}\boldsymbol{A}^2t^2+\cdots+\frac{1}{(k-1)!}\boldsymbol{A}^kt^{k-1}+\cdots=$$
$$\boldsymbol{A}\cdot\mathrm{e}^{\boldsymbol{A}t}=\mathrm{e}^{\boldsymbol{A}t}\cdot\boldsymbol{A}=\boldsymbol{A}\boldsymbol{\Phi}(t)=\boldsymbol{\Phi}(t)\boldsymbol{A}$$

(3) $\boldsymbol{\Phi}(t\pm\tau)=\boldsymbol{\Phi}(t)\boldsymbol{\Phi}(\pm\tau)=\boldsymbol{\Phi}(\pm\tau)\boldsymbol{\Phi}(t)$ (2.12)

证明 $\boldsymbol{\Phi}(t\pm\tau)=\mathrm{e}^{\boldsymbol{A}(t\pm\tau)}=\mathrm{e}^{\boldsymbol{A}t}\cdot\mathrm{e}^{\pm\boldsymbol{A}\tau}=\mathrm{e}^{\pm\boldsymbol{A}\tau}\cdot\mathrm{e}^{\boldsymbol{A}t}=\boldsymbol{\Phi}(t)\boldsymbol{\Phi}(\pm\tau)=\boldsymbol{\Phi}(\pm\tau)\boldsymbol{\Phi}(t)$

(4) $\boldsymbol{\Phi}^{-1}(t)=\boldsymbol{\Phi}(-t)$ (2.13)

证明 $\boldsymbol{\Phi}^{-1}(t)=(\mathrm{e}^{\boldsymbol{A}t})^{-1}=\mathrm{e}^{-\boldsymbol{A}t}=\boldsymbol{\Phi}(-t)$

(5) $\boldsymbol{\Phi}(t_2-t_0)=\boldsymbol{\Phi}(t_2-t_1)\boldsymbol{\Phi}(t_1-t_0)$ (2.14)

证明 $\boldsymbol{\Phi}(t_2-t_1)\boldsymbol{\Phi}(t_1-t_0)=\mathrm{e}^{\boldsymbol{A}(t_2-t_1)}\mathrm{e}^{\boldsymbol{A}(t_1-t_0)}=\mathrm{e}^{\boldsymbol{A}(t_2-t_0)}=\boldsymbol{\Phi}(t_2-t_0)$

(6) $[\boldsymbol{\Phi}(t)]^k=\boldsymbol{\Phi}(kt)$,$k$ 为正整数 (2.15)

证明 $[\boldsymbol{\Phi}(t)]^k=(\mathrm{e}^{\boldsymbol{A}t})^k=\mathrm{e}^{k\boldsymbol{A}t}=\boldsymbol{\Phi}(kt)$

(7) $\mathrm{e}^{(\boldsymbol{A}+\boldsymbol{B})t}=\mathrm{e}^{\boldsymbol{A}t}\cdot\mathrm{e}^{\boldsymbol{B}t}$ 的充分必要条件是 $\boldsymbol{AB}=\boldsymbol{BA}$ (2.16)

证明 由于

$$\mathrm{e}^{(\boldsymbol{A}+\boldsymbol{B})t}=\boldsymbol{I}_n+(\boldsymbol{A}+\boldsymbol{B})t+\frac{1}{2!}(\boldsymbol{A}+\boldsymbol{B})^2t^2+\frac{1}{3!}(\boldsymbol{A}+\boldsymbol{B})^3t^3+\cdots=$$

$$I_n + (A + B)t + \frac{1}{2!}(A^2 + AB + BA + B^2)t^2 + \frac{1}{3!}(A^3 + A^2B + ABA + AB^2 + BA^2 + BAB + B^2A + B^3)t^3 + \cdots$$

而且

$$e^{At} \cdot e^{Bt} = \left(I_n + At + \frac{1}{2!}A^2t^2 + \frac{1}{3!}A^3t^3 + \cdots\right)\left(I_n + Bt + \frac{1}{2!}B^2t^2 + \frac{1}{3!}B^3t^3 + \cdots\right) = I_n + (A + B)t + \frac{1}{2!}(A^2 + 2AB + B^2)t^2 + \frac{1}{3!}(A^3 + 3A^2B + 3AB^2 + B^3)t^3 + \cdots$$

比较这两个式子得到

$$e^{(A+B)t} - e^{At}e^{Bt} = \frac{1}{2!}(BA - AB)^2t^2 + \frac{1}{3!}(-2A^2B + ABA - 2AB^2 + BA^2 + BAB + B^2A)t^3 + \cdots$$

由结果所得到的公式可以看到，当且仅当 $AB = BA$ 时

$$e^{(A+B)t} = e^{At} \cdot e^{Bt}$$

4. $x(t_0) = x_0$ 时的齐次解

现在推导当初始状态定义在某一个一般时间 $t_0 \neq 0$ 时，即 $x(t_0) = x_0$ 时的齐次解。由于

$$x(t_0) = \Phi(t_0)x(0)$$

因此

$$x(t) = \Phi(t)x(0) = \Phi(t)\Phi^{-1}(t_0)x(t_0) = \Phi(t)\Phi(-t_0)x(t_0) = \Phi(t - t_0)x(t_0) = \Phi(t - t_0)x_0 = e^{A(t-t_0)}x(t_0) = e^{A(t-t_0)}x_0 \tag{2.17}$$

2.1.2　e^{At} 的估算

1. 用 e^{At} 的无穷级数

矩阵指数函数 e^{At} 可以用式(2.7) 中的无穷级数估算

$$e^{At} \approx \sum_{k=0}^{N} \frac{1}{k!}A^kt^k \tag{2.18}$$

其中 N 是一个有限大整数。一般地讲，这一级数将给不出一个闭式解，它主要用于计算机估算。

2. 用拉普拉斯变换

对式(2.1) 取拉普拉斯变换得到

$$sX(s) - x(0) = AX(s)$$

或

$$X(s) = (sI - A)^{-1}x(0)$$

将该式与式(2.8) 中的解相比较，有

$$e^{At} = \mathscr{L}^{-1}[(sI - A)^{-1}] \tag{2.19}$$

例 2.1　已知矩阵 $\boldsymbol{A} = \begin{bmatrix} 0 & 1 \\ -2 & -3 \end{bmatrix}$,计算矩阵指数函数 e^{At}。

解　为了找出矩阵指数函数,首先计算矩阵$(s\boldsymbol{I} - \boldsymbol{A})^{-1}$

$$(s\boldsymbol{I} - \boldsymbol{A})^{-1} = \begin{bmatrix} s & -1 \\ 2 & s+3 \end{bmatrix}^{-1} = \frac{\mathrm{adj}(s\boldsymbol{I} - \boldsymbol{A})}{\det(s\boldsymbol{I} - \boldsymbol{A})} = \frac{1}{(s+1)(s+2)}\begin{bmatrix} s+3 & 1 \\ -2 & s \end{bmatrix} =$$

$$\begin{bmatrix} \dfrac{2}{s+1} - \dfrac{1}{s+2} & \dfrac{1}{s+1} - \dfrac{1}{s+2} \\ -\dfrac{2}{s+1} + \dfrac{2}{s+2} & -\dfrac{1}{s+1} + \dfrac{2}{s+2} \end{bmatrix}$$

矩阵指数是$(s\boldsymbol{I} - \boldsymbol{A})^{-1}$的拉普拉斯反变换,即

$$e^{At} = \mathscr{L}^{-1}[(s\boldsymbol{I} - \boldsymbol{A})^{-1}] = \begin{bmatrix} 2e^{-t} - e^{-2t} & e^{-t} - e^{-2t} \\ -2e^{-t} + 2e^{-2t} & -e^{-t} + 2e^{-2t} \end{bmatrix}$$

3. 用约当规范形

由于矩阵的指数函数用约当规范形来表示时很简单,而通过相似变换可以使任意一个给定的系统矩阵变换为约当形,因此相似变换是一种计算 e^{At} 的重要方法。

(1) 约当形的指数矩阵

情况 1　对角阵的矩阵指数

如果一个 $n \times n$ 矩阵是对角阵,即

$$\boldsymbol{\Lambda} = \begin{bmatrix} \lambda_1 & & & \\ & \lambda_2 & & \\ & & \ddots & \\ & & & \lambda_n \end{bmatrix} \tag{2.20}$$

则其矩阵指数仍然是一个 $n \times n$ 的对角阵,并由下式给出为

$$e^{\Lambda t} = \begin{bmatrix} e^{\lambda_1 t} & & & \\ & e^{\lambda_2 t} & & \\ & & \ddots & \\ & & & e^{\lambda_n t} \end{bmatrix} \tag{2.21}$$

情况 2　一般约当阵的矩阵指数

假设一个 $n \times n$ 约当阵 $\boldsymbol{J}$ 有 m 个互异的特征根$\lambda_1, \cdots, \lambda_i, \cdots, \lambda_m$,而且特征根 λ_i 的重数为 σ_i,即

$$\boldsymbol{J} = \begin{bmatrix} \boldsymbol{J}_1 & \cdots & 0 & \cdots & 0 \\ \vdots & & \vdots & & \vdots \\ 0 & \cdots & \boldsymbol{J}_i & \cdots & 0 \\ \vdots & & \vdots & & \vdots \\ 0 & \cdots & 0 & \cdots & \boldsymbol{J}_m \end{bmatrix}_{n \times n} \tag{2.22}$$

而约当块

$$J_i = \begin{bmatrix} \lambda_i & 1 & \cdots & 0 & 0 \\ 0 & \lambda_i & \cdots & 0 & 0 \\ \vdots & \vdots & & \vdots & \vdots \\ 0 & 0 & \cdots & \lambda_i & 1 \\ 0 & 0 & \cdots & 0 & \lambda_i \end{bmatrix}_{\sigma_i \times \sigma_i} \tag{2.23}$$

是一个 $\sigma_i \times \sigma_i$ 的方块矩阵。这样，相应的矩阵指数由下式给出

$$e^{Jt} = \begin{bmatrix} e^{J_1 t} & \cdots & 0 & \cdots & 0 \\ \vdots & & \vdots & & \vdots \\ 0 & \cdots & e^{J_i t} & \cdots & 0 \\ \vdots & & \vdots & & \vdots \\ 0 & \cdots & 0 & \cdots & e^{J_m t} \end{bmatrix} \tag{2.24}$$

其中

$$e^{J_i t} = I + J_i t + \frac{1}{2!}J_i^2 t^2 + \cdots = \begin{bmatrix} 1 & t & \frac{1}{2!}t^2 & \cdots & \frac{1}{(\sigma_i - 1)!}t^{\sigma_i - 1} \\ 0 & 1 & t & \cdots & \frac{1}{(\sigma_i - 2)!}t^{\sigma_i - 2} \\ 0 & 0 & 1 & \cdots & \frac{1}{(\sigma_i - 3)!}t^{\sigma_i - 3} \\ \vdots & \vdots & \vdots & & \vdots \\ 0 & 0 & 0 & \cdots & 1 \end{bmatrix} e^{\lambda_i t} \tag{2.25}$$

利用矩阵指数函数的幂级数展开很容易证明式(2.20) 和式(2.25)。

(2) 利用约当形计算 e^{At}

由矩阵理论，对于给定矩阵 A 存在非奇异常数阵 P 使得 $\bar{A} = P^{-1}AP$。这样就有

$$e^{\bar{A}t} = e^{(P^{-1}AP)t} = I + P^{-1}APt + \frac{1}{2!}(P^{-1}AP)^2 t^2 + \cdots + \frac{1}{k!}(P^{-1}AP)^k t^k + \cdots =$$

$$P^{-1}IP + P^{-1}APt + \frac{1}{2!}P^{-1}A^2Pt^2 + \cdots + \frac{1}{k!}P^{-1}A^kPt^k + \cdots =$$

$$P^{-1}[I + P^{-1}APt + \frac{1}{2!}(P^{-1}AP)^2t^2 + \cdots]P = P^{-1}e^{At}P$$

由此，系统矩阵 A 的矩阵指数函数由下式给出

$$e^{At} = Pe^{(P^{-1}AP)t}P^{-1} \tag{2.26}$$

尤其是，如果 $P^{-1}AP$ 为约当形包括对角形时，由式(2.21)、式(2.24) 和式(2.25) 很容易获得矩阵指数 e^{At}。

(3) 变换方阵为约当形

情况 1　矩阵 A 的所有特征值互异

假设 $n \times n$ 矩阵 A 的所有特征值 $\lambda_1, \lambda_2, \cdots, \lambda_n$ 互异。在这种情况下，存在相似变换 $P^{-1}AP = \Lambda$ 使得矩阵 Λ 为对角形。

矩阵 P 可以由矩阵 A 的线性无关特征向量组成，即

$$P = [p_1 \quad p_2 \quad \cdots \quad p_n] \tag{2.27}$$

其中 $p_i(i = 1,2,\cdots,n)$ 表示与特征值 λ_i 相关的特征向量。任何满足 $Ap = \lambda p$ 的非零向量 p 都称为矩阵 A 的与特征值 λ 相关的特征向量。与特征值 λ_i 相关的特征向量 p_i 由下式给出

$$(\lambda_i I - A)p_i = 0,\ i = 1,2,\cdots,n \tag{2.28}$$

特别是当 A 为伴随矩阵时,可以证明矩阵 P 可以是一个范德蒙阵,即

$$P = \begin{bmatrix} 1 & 1 & \cdots & 1 \\ \lambda_1 & \lambda_2 & \cdots & \lambda_n \\ \vdots & \vdots & & \vdots \\ \lambda_1^{n-1} & \lambda_2^{n-1} & \cdots & \lambda_n^{n-1} \end{bmatrix} \tag{2.29}$$

其中 $\lambda_1,\lambda_2,\cdots,\lambda_n$ 为矩阵 A 的特征值。

例 2.2　仍然考虑例 2.1 中给出的例子,即

$$A = \begin{bmatrix} 0 & 1 \\ -2 & -3 \end{bmatrix}$$

用相似变换计算矩阵指数函数。

解　由于

$$\det(sI - A) = \begin{vmatrix} s & -1 \\ 2 & s+3 \end{vmatrix} = s^2 + 3s + 2 = (s+1)(s+2)$$

特征值为 $\lambda_1 = -1$ 和 $\lambda_2 = -2$。由观察可知,矩阵 A 是一个伴随矩阵,进行相似变换时可以用范德蒙阵,即

$$P = \begin{bmatrix} 1 & 1 \\ \lambda_1 & \lambda_2 \end{bmatrix} = \begin{bmatrix} 1 & 1 \\ -1 & -2 \end{bmatrix}$$

P 的逆矩阵为

$$P^{-1} = \frac{\operatorname{adj} P}{\det P} = \begin{bmatrix} 2 & 1 \\ -1 & -1 \end{bmatrix}$$

因此,现在可以获得对角形

$$\Lambda = P^{-1}AP = \begin{bmatrix} -1 & 0 \\ 0 & -2 \end{bmatrix}$$

最后,由式(2.26)知,矩阵指数函数为

$$e^{AT} = Pe^{(P^{-1}AP)t}P^{-1} = Pe^{\Lambda t}P^{-1} = \begin{bmatrix} 1 & 1 \\ -1 & -2 \end{bmatrix}\begin{bmatrix} e^{-t} & 0 \\ 0 & e^{-2t} \end{bmatrix}\begin{bmatrix} 2 & 1 \\ -1 & -1 \end{bmatrix} =$$

$$\begin{bmatrix} 2e^{-t} - e^{-2t} & e^{-t} - e^{-2t} \\ -2e^{-t} + 2e^{-2t} & -e^{-t} + 2e^{-2t} \end{bmatrix}$$

情况 2　矩阵 A 的特征值不全互异

当矩阵 A 有重特征值时,如果与各重特征值相关的独立特征向量数都等于该特征值的重数,则矩阵 A 可以对角化。

例 2.3　将以下矩阵变换为对角阵

$$A = \begin{bmatrix} 1 & 0 & -1 \\ 0 & 1 & 0 \\ 0 & 0 & 2 \end{bmatrix}$$

解　A 的特征值为$\lambda_1 = \lambda_2 = 1, \lambda_3 = 2$。与 $\lambda = 1$ 相关的特征向量可以通过求解以下方程获取

$$(\lambda_1 I - A)p = \begin{bmatrix} 0 & 0 & -1 \\ 0 & 0 & 0 \\ 0 & 0 & -1 \end{bmatrix} p = 0$$

注意,对于这一方程可以找到两个独立的,与 $\lambda_1 = \lambda_2 = 1$ 相关的特征向量,譬如 $p_1 = [1 \quad 0 \quad 0]^T$ 和 $p_2 = [0 \quad 1 \quad 0]^T$。求解方程

$$(\lambda_3 I - A)p = \begin{bmatrix} 1 & 0 & 1 \\ 0 & 1 & 0 \\ 0 & 0 & 0 \end{bmatrix} p = 0$$

可以找到与 $\lambda_3 = 2$ 相关的特征向量,如 $p_3 = [1 \quad 0 \quad -1]^T$。因此得

$$P = [p_1 \quad p_2 \quad p_3] = \begin{bmatrix} 1 & 0 & 1 \\ 0 & 1 & 0 \\ 0 & 0 & -1 \end{bmatrix}, \ P^{-1} = \frac{\mathrm{adj} P}{\det P} = \begin{bmatrix} 1 & 0 & 1 \\ 0 & 1 & 0 \\ 0 & 0 & -1 \end{bmatrix}$$

由此可以估算矩阵 $P^{-1}AP$,并确定为

$$P^{-1}AP = \begin{bmatrix} 1 & 0 & 0 \\ 0 & 1 & 0 \\ 0 & 0 & 2 \end{bmatrix} = \Lambda$$

在这一例子中,具有重特征值的矩阵可以对角化。但是,一般情况下,含有重特征值的矩阵并不总是可以对角化的。这时应用相似变换只能使矩阵 A 变换为约当形,即

$$P^{-1}AP = J$$

当一个矩阵 A 具有重特征值时,与重特征值相关的特征向量数常常少于特征值的重数,因此组成变换阵还需要更多的向量。这样的一些向量与相应的特征向量有关,并称为广义特征向量。例如,如果矩阵 A 有$\lambda_1 = \lambda_2 = \lambda_3$,而且与该特征值相关的特征向量只有一个并记为 p_1,那么两个广义特征向量记为 p_2 和 p_3,并可以由以下方程获取

$$(\lambda_1 I - A)p_2 = -p_1$$
$$(\lambda_1 I - A)p_3 = -p_2$$

例 2.4　将以下矩阵变换为约当形

$$A = \begin{bmatrix} 1 & 1 & 2 \\ 0 & 1 & 3 \\ 0 & 0 & 2 \end{bmatrix}$$

解　A 的特征值为$\lambda_1 = \lambda_2 = 1$ 和 $\lambda_3 = 2$。与 $\lambda = 1$ 相关的特征向量可以通过求解以下方程找到

$$(\lambda_1 I - A)p = \begin{bmatrix} 0 & -1 & -2 \\ 0 & 0 & -3 \\ 0 & 0 & -1 \end{bmatrix} p = 0$$

可以看到,只能找到一个与 $\lambda_1 = \lambda_2 = 1$ 相关的线性独立特征向量,如 $\boldsymbol{p}_1 = [1\quad 0\quad 0]^{\mathrm{T}}$。组成变换阵所需的另一个与 $\lambda_1 = \lambda_2 = 1$ 相关的向量可以通过求解以下方程找到

$$(\lambda_1 \boldsymbol{I} - \boldsymbol{A})\boldsymbol{p}_2 = -\boldsymbol{p}_1$$

即

$$\begin{bmatrix} 0 & -1 & -2 \\ 0 & 0 & -3 \\ 0 & 0 & -1 \end{bmatrix} \boldsymbol{p}_2 = -\begin{bmatrix} 1 \\ 0 \\ 0 \end{bmatrix}$$

显然,$\boldsymbol{p}_2 = [0\quad 1\quad 0]^{\mathrm{T}}$ 是一个向量解。求解方程

$$(\lambda_3 \boldsymbol{I} - \boldsymbol{A})\boldsymbol{p}_3 = \boldsymbol{0}$$

得到与 $\lambda_3 = 2$ 相关的特征向量,譬如 $\boldsymbol{p}_3 = [5\quad 3\quad 1]^{\mathrm{T}}$。因此有

$$\boldsymbol{P} = \begin{bmatrix} 1 & 0 & 5 \\ 0 & 1 & 3 \\ 0 & 0 & 1 \end{bmatrix},\ \boldsymbol{P}^{-1} = \begin{bmatrix} 1 & 0 & -5 \\ 0 & 1 & -3 \\ 0 & 0 & 1 \end{bmatrix}$$

变换的结果为

$$\boldsymbol{P}^{-1}\boldsymbol{A}\boldsymbol{P} = \begin{bmatrix} 1 & 1 & 0 \\ 0 & 1 & 0 \\ 0 & 0 & 2 \end{bmatrix} = \boldsymbol{J}$$

2.1.3　非齐次方程的解

设状态方程为

$$\dot{\boldsymbol{x}}(t) = \boldsymbol{A}\boldsymbol{x}(t) + \boldsymbol{B}\boldsymbol{u}(t),\ t \geqslant 0;\ \boldsymbol{x}(0) = \boldsymbol{x}_0 \tag{2.30}$$

我们现在考虑该方程解的一般形式。该状态方程的全解可以用两种方法分别导出。

1.积分法

由式(2.30)得到

$$\mathrm{e}^{-\boldsymbol{A}t}(\dot{\boldsymbol{x}} - \boldsymbol{A}\boldsymbol{x}) = \mathrm{e}^{-\boldsymbol{A}t}\boldsymbol{B}\boldsymbol{u}$$

由于

$$\frac{\mathrm{d}}{\mathrm{d}t}[\mathrm{e}^{-\boldsymbol{A}t}\boldsymbol{x}] = \mathrm{e}^{-\boldsymbol{A}t}\dot{\boldsymbol{x}} + \mathrm{e}^{-\boldsymbol{A}t}(-\boldsymbol{A})\boldsymbol{x} = \mathrm{e}^{-\boldsymbol{A}t}[\dot{\boldsymbol{x}} - \boldsymbol{A}\boldsymbol{x}]$$

因此

$$\frac{\mathrm{d}}{\mathrm{d}t}[\mathrm{e}^{-\boldsymbol{A}t}\boldsymbol{x}] = \mathrm{e}^{-\boldsymbol{A}t}\boldsymbol{B}\boldsymbol{u}$$

对这一方程进行积分得到

$$\mathrm{e}^{-\boldsymbol{A}t}\boldsymbol{x}(t) - \boldsymbol{x}(0) = \int_0^t \mathrm{e}^{-\boldsymbol{A}\tau}\boldsymbol{B}\boldsymbol{u}(\tau)\mathrm{d}\tau$$

全解由下式给出

$$\boldsymbol{x}(t) = \mathrm{e}^{\boldsymbol{A}t}\boldsymbol{x}(0) + \int_0^t \mathrm{e}^{\boldsymbol{A}(t-\tau)}\boldsymbol{B}\boldsymbol{u}(\tau)\mathrm{d}\tau =$$

$$\boldsymbol{\Phi}(t)\boldsymbol{x}(0) + \int_0^t \boldsymbol{\Phi}(t-\tau)\boldsymbol{B}\boldsymbol{u}(\tau)\mathrm{d}\tau \tag{2.31}$$

如果 $t_0 \neq 0$ 则全解由下式给出

$$\boldsymbol{x}(t) = \mathrm{e}^{\boldsymbol{A}(t-t_0)}\boldsymbol{x}(t_0) + \int_{t_0}^{t} \mathrm{e}^{\boldsymbol{A}(t-\tau)}\boldsymbol{B}\boldsymbol{u}(\tau)\mathrm{d}\tau =$$

$$\boldsymbol{\Phi}(t - t_0)\boldsymbol{x}(t_0) + \int_{t_0}^{t}\boldsymbol{\Phi}(t-\tau)\boldsymbol{B}\boldsymbol{u}(\tau)\mathrm{d}\tau \tag{2.32}$$

2.拉普拉斯变换法

仍然考虑状态方程

$$\dot{\boldsymbol{x}}(t) = \boldsymbol{A}\boldsymbol{x}(t) + \boldsymbol{B}\boldsymbol{u}(t),\ t \geqslant 0$$

取拉普拉斯变换得到

$$s\boldsymbol{X}(s) - \boldsymbol{x}(0) = \boldsymbol{A}\boldsymbol{X}(s) + \boldsymbol{B}\boldsymbol{u}(s)$$

或

$$(s\boldsymbol{I} - \boldsymbol{A})\boldsymbol{X}(s) = \boldsymbol{x}(0) + \boldsymbol{B}\boldsymbol{u}(s)$$

因此

$$\boldsymbol{x}(t) = \mathscr{L}^{-1}[(s\boldsymbol{I}-\boldsymbol{A})^{-1}]\boldsymbol{x}(0) + \mathscr{L}^{-1}[(s\boldsymbol{I}-\boldsymbol{A})^{-1}\boldsymbol{B}\boldsymbol{u}(s)] \tag{2.33}$$

例 2.5　求解下式给出的状态方程

$$\dot{\boldsymbol{x}}(t) = \begin{bmatrix} 0 & 1 \\ -2 & -3 \end{bmatrix}\boldsymbol{x}(t) + \begin{bmatrix} 0 \\ 1 \end{bmatrix}\boldsymbol{u}(t),\ \boldsymbol{u}(t) = 1(t),\ \boldsymbol{x}(0) = \begin{bmatrix} x_1(0) \\ x_2(0) \end{bmatrix}$$

解　(1) 用积分法。由于 $\boldsymbol{u}(t) = 1(t)$,令 $\xi = t - \tau$ 得到

$$\boldsymbol{x}(t) = \boldsymbol{\Phi}(t)\boldsymbol{x}(0) + \int_0^t \boldsymbol{\Phi}(t-\tau)\boldsymbol{B}\boldsymbol{u}(\tau)\mathrm{d}\tau = \boldsymbol{\Phi}(t)\boldsymbol{x}(0) - \int_t^0 \boldsymbol{\Phi}(\xi)\boldsymbol{B}\mathrm{d}\xi =$$

$$\boldsymbol{\Phi}(t)\boldsymbol{x}(0) + \int_0^t \boldsymbol{\Phi}(\tau)\boldsymbol{B}\mathrm{d}\tau$$

由例 2.1 有

$$\boldsymbol{\Phi}(t) = \begin{bmatrix} 2\mathrm{e}^{-t} - \mathrm{e}^{-2t} & \mathrm{e}^{-t} - \mathrm{e}^{-2t} \\ -2\mathrm{e}^{-t} + 2\mathrm{e}^{-2t} & -\mathrm{e}^{-t} + 2\mathrm{e}^{-2t} \end{bmatrix}$$

因此

$$\int_0^t \boldsymbol{\Phi}(\tau)\boldsymbol{B}\mathrm{d}\tau = \int_0^t \begin{bmatrix} 2\mathrm{e}^{-\tau} - \mathrm{e}^{-2\tau} & \mathrm{e}^{-\tau} - \mathrm{e}^{-2\tau} \\ -2\mathrm{e}^{-\tau} + 2\mathrm{e}^{-2\tau} & -\mathrm{e}^{-\tau} + 2\mathrm{e}^{-2\tau} \end{bmatrix}\begin{bmatrix} 0 \\ 1 \end{bmatrix}\mathrm{d}\tau = \int_0^t \begin{bmatrix} \mathrm{e}^{-\tau} - \mathrm{e}^{-2\tau} \\ -\mathrm{e}^{-\tau} + 2\mathrm{e}^{-2\tau} \end{bmatrix}\mathrm{d}\tau =$$

$$\left.\begin{bmatrix} -\mathrm{e}^{-\tau} + \dfrac{1}{2}\mathrm{e}^{-2\tau} \\ \mathrm{e}^{-\tau} - \mathrm{e}^{-2\tau} \end{bmatrix}\right|_0^t = \begin{bmatrix} \dfrac{1}{2} - \mathrm{e}^{-t} + \dfrac{1}{2}\mathrm{e}^{-2t} \\ \mathrm{e}^{-t} - \mathrm{e}^{-2t} \end{bmatrix}$$

$$\boldsymbol{x}(t) = \begin{bmatrix} 2\mathrm{e}^{-t} - \mathrm{e}^{-2t} & \mathrm{e}^{-t} - \mathrm{e}^{-2t} \\ -2\mathrm{e}^{-t} + 2\mathrm{e}^{-2t} & -\mathrm{e}^{-t} + 2\mathrm{e}^{-2t} \end{bmatrix}\begin{bmatrix} x_1(0) \\ x_2(0) \end{bmatrix} + \begin{bmatrix} \dfrac{1}{2} - \mathrm{e}^{-t} + \dfrac{1}{2}\mathrm{e}^{-2t} \\ \mathrm{e}^{-t} - \mathrm{e}^{-2t} \end{bmatrix}$$

(2) 用拉普拉斯变换法。由例 2.1 有

$$(s\boldsymbol{I} - \boldsymbol{A})^{-1} = \begin{bmatrix} \dfrac{2}{s+1} - \dfrac{1}{s+2} & \dfrac{1}{s+1} - \dfrac{1}{s+2} \\ -\dfrac{2}{s+1} + \dfrac{2}{s+2} & -\dfrac{1}{s+1} + \dfrac{2}{s+2} \end{bmatrix}$$

$$\mathscr{L}^{-1}[(s\boldsymbol{I}-\boldsymbol{A})^{-1}]=\begin{bmatrix}2\mathrm{e}^{-t}-\mathrm{e}^{-2t} & \mathrm{e}^{-t}-\mathrm{e}^{-2t}\\ -2\mathrm{e}^{-t}+2\mathrm{e}^{-2t} & -\mathrm{e}^{-t}+2\mathrm{e}^{-2t}\end{bmatrix}$$

由此

$$(s\boldsymbol{I}-\boldsymbol{A})^{-1}\boldsymbol{B}U(s)=\begin{bmatrix}\frac{2}{s+1}-\frac{1}{s+2} & \frac{1}{s+1}-\frac{1}{s+2}\\ -\frac{2}{s+1}+\frac{2}{s+2} & -\frac{1}{s+1}+\frac{2}{s+2}\end{bmatrix}\begin{bmatrix}0\\1\end{bmatrix}\cdot\frac{1}{s}=$$

$$\begin{bmatrix}\frac{1}{s+1}-\frac{1}{s+2}\\ -\frac{1}{s+1}+\frac{2}{s+2}\end{bmatrix}\cdot\frac{1}{s}=\begin{bmatrix}\frac{1}{s(s+1)(s+2)}\\ \frac{1}{(s+1)(s+2)}\end{bmatrix}=$$

$$\begin{bmatrix}\frac{1}{2}\cdot\frac{1}{s}-\frac{1}{s+1}+\frac{1}{2}\cdot\frac{1}{s+2}\\ \frac{1}{s+1}-\frac{1}{s+2}\end{bmatrix}$$

由式(2.33)有

$$\boldsymbol{x}(t)=\mathscr{L}^{-1}[(s\boldsymbol{I}-\boldsymbol{A})^{-1}]\boldsymbol{x}(0)+\mathscr{L}^{-1}[(s\boldsymbol{I}-\boldsymbol{A})^{-1}\boldsymbol{B}\boldsymbol{u}(s)]=$$

$$\begin{bmatrix}2\mathrm{e}^{-t}-\mathrm{e}^{-2t} & \mathrm{e}^{-t}-\mathrm{e}^{-2t}\\ -2\mathrm{e}^{-t}+2\mathrm{e}^{-2t} & -\mathrm{e}^{-t}+2\mathrm{e}^{-2t}\end{bmatrix}\begin{bmatrix}x_1(0)\\x_2(0)\end{bmatrix}+\begin{bmatrix}\frac{1}{2}-\mathrm{e}^{-t}+\frac{1}{2}\mathrm{e}^{-2t}\\ \mathrm{e}^{-t}-\mathrm{e}^{-2t}\end{bmatrix}$$

2.2 时不变离散系统

2.2.1 状态空间表达式

线性离散时间系统通常采用差分方程或者脉冲传递函数描述。单变量系统差分方程的一般形式为

$$\begin{aligned}&y(k+n)+a_1y(k+n-1)+\cdots+a_{n-1}y(k+1)+a_ny(k)=\\&b_1u(k+n-1)+\cdots+b_{n-1}u(k+1)+b_nu(k)\end{aligned}\tag{2.34}$$

其中 $u(k)$ 和 $y(k)$ 分别为第 k 个采样时刻的输入和输出。由连续时间系统相似的方法，可以建立相应的规范形状态空间表达式。

例如，如果状态变量选择为

$$\begin{aligned}x_1(k)&=z(k)\\x_2(k)&=z(k+1)\\&\vdots\\x_n(k)&=z(k+n-1)\end{aligned}$$

并使得方程 $z(k+n)+a_1z(k+n-1)+\cdots+a_nz(k)=u(k)$ 成立。由此得到的状态空间表达式为能控规范形

$$\left.\begin{aligned} x(k+1) &= \begin{bmatrix} 0 & \vdots & \\ \vdots & \vdots & I_{n-1} \\ 0 & \vdots & \\ \cdots & \cdots & \cdots \\ -a_n & -a_{n-1} & \cdots \quad -a_1 \end{bmatrix} x(k) + \begin{bmatrix} 0 \\ \vdots \\ 0 \\ 1 \end{bmatrix} u(k) \\ y(k) &= [b_n \quad b_{n-1} \quad \cdots \quad b_1] x(k) \end{aligned}\right\} \tag{2.35}$$

离散系统状态空间表达式的标准形式记为

$$\left.\begin{aligned} x(k+1) &= Ax(k) + Bu(k) \\ y(k) &= Cx(k) + Du(k) \end{aligned}\right\} \tag{2.36}$$

其中矩阵 D 通常为零。注意，离散时间状态方程是一组一阶差分方程而不是像连续时间情况下那样的一组微分方程。

2.2.2 连续时间状态方程的离散化

如果一个连续时间系统只对一定时刻上的响应感兴趣或者进行测量，该系统也可作为离散时间系统建立模型。为了方便起见，假设输出出现的离散时刻是等间隔的，间隔时间记为 T。

1.离散化

线性时不变连续系统的状态方程通常写成

$$\dot{x}(t) = Ax(t) + Bu(t)$$

该方程的解则为

$$x(t) = e^{A(t-t_0)}x(t_0) + \int_{t_0}^{t} e^{A(t-\tau)}Bu(\tau)d\tau$$

取 $t_0 = kT$ 和 $t = (k+1)T$ 则有

$$x[(k+1)T] = e^{AT}x(kT) + \int_{kT}^{(k+1)T} e^{A[(k+1)T-\tau]}Bu(\tau)d\tau$$

令 $\tau = kT + \xi$，其中 $0 \leqslant \zeta \leqslant T$，将其代入上面的方程得到

$$x[(k+1)T] = e^{AT}x(kT) + \int_0^T e^{A(T-\xi)}Bu(kT+\xi)d\xi \approx$$

$$e^{AT}x(kT) + \int_0^T e^{A(T-\epsilon)}Bd\xi u(kT)$$

或

$$x[(k+1)T] = A^* x(kT) + B^* u(kT) \tag{2.37}$$

式中

$$A^* = e^{AT} = e^{At}|_{t=T} \tag{2.38}$$

$$B^* = \int_0^T e^{A(T-\xi)}Bd\xi = -\int_T^0 e^{At}Bdt = \int_0^T e^{At}Bdt \tag{2.39}$$

作为代数方程，离散后的输出方程则为

$$y(k) = Cx(k) + Du(k) \tag{2.40}$$

例 2.6　确定下面连续时间状态方程的离散化方程

$$\dot{x}(t)=\begin{bmatrix}0 & 1\\ -2 & -3\end{bmatrix}x(t)+\begin{bmatrix}0\\1\end{bmatrix}u(t)$$

假设间隔时间 $T=1$ s。

解 由例 2.1 有

$$e^{At}=\begin{bmatrix}2e^{-t}-e^{-2t} & e^{-t}-e^{-2t}\\ -2e^{-t}+2e^{-2t} & -e^{-t}+2e^{-2t}\end{bmatrix}$$

应用式(2.38) 和式(2.39) 得到

$$A^*=e^{At}\big|_{t=T}=\begin{bmatrix}2e^{-t}-e^{-2t} & e^{-t}-e^{-2t}\\ -2e^{-t}+2e^{-2t} & -e^{-t}+2e^{-2t}\end{bmatrix}\Bigg|_{t=1}=\begin{bmatrix}0.600\,4 & 0.232\,5\\ -0.465\,1 & -0.097\,2\end{bmatrix}$$

$$B^*=\int_0^T e^{At}B\,dt=\int_0^1\begin{bmatrix}2e^{-t}-e^{-2t} & e^{-t}-e^{-2t}\\ -2e^{-t}+2e^{-2t} & -e^{-t}+2e^{-2t}\end{bmatrix}\begin{bmatrix}0\\1\end{bmatrix}dt=\int_0^1\begin{bmatrix}e^{-t}-e^{-2t}\\ -e^{-t}+2e^{-2t}\end{bmatrix}dt=$$

$$\begin{bmatrix}\frac{1}{2}-e^{-1}+\frac{1}{2}e^{-2}\\ e^{-1}-2e^{-2}\end{bmatrix}=\begin{bmatrix}0.199\,8\\0.232\,5\end{bmatrix}$$

最终离散化后的状态方程为

$$x(k+1)=\begin{bmatrix}0.600\,4 & 0.232\,5\\ -0.465\,1 & -0.097\,2\end{bmatrix}x(k)+\begin{bmatrix}0.199\,8\\0.232\,5\end{bmatrix}u(k)$$

2.近似表达式

如果与系统的时间常数相比采样周期 T 足够小,则有

$$A^*=e^{AT}=I+AT+\cdots\approx I+AT$$

和

$$B^*=\int_0^T e^{At}B\,dt=\int_0^T e^{At}dt\cdot B\approx\int_0^T(I+At)dt\cdot B\approx TB$$

由此,式(2.37) 就可以近似地改写成

$$x[(k+1)T]=(I+TA)x(kT)+TBu(kT)\tag{2.41}$$

2.2.3 离散时间状态方程的解

离散时间状态方程的解可以用经典法、z 变换法,或者迭代法获取。这里随后将用实例说明迭代法。

基本的离散时间状态方程为

$$x(k+1)=Ax(k)+Bu(k)$$

令 $k=0,1,\cdots,k-1$ 得到

$$x(1)=Ax(0)+Bu(0)$$
$$x(2)=Ax(1)+Bu(1)=A^2x(0)+ABu(0)+Bu(1)$$
$$x(3)=Ax(2)+Bu(2)=A^3x(0)+A^2Bu(0)+ABu(1)+Bu(2)$$
$$\vdots$$
$$x(k)=Ax(k-1)+Bu(k-1)=A^kx(0)+A^{k-1}Bu(0)+\cdots+ABu(k-2)+Bu(k-1)$$

这样,状态方程的解由迭代方程给出为

$$x(k) = A^k x(0) + \sum_{i=0}^{k-1} A^{k-i-1} Bu(i) \tag{2.42}$$

在理论研究中,式(2.42) 经常写成

$$x(k) = \Phi(k)x(0) + \sum_{i=0}^{k-1} \Phi(k-i-1)Bu(i) = \Phi(k)x(0) + \sum_{i=0}^{k-1} \Phi(i)Bu(k-i-1) \tag{2.43}$$

其中 $\Phi(k) = A^k$ 称为离散状态转移矩阵。

习　题

题 2.1　计算以下矩阵的预解矩阵,即$(sI - A)^{-1}$和状态转移矩阵

(a) $A = \begin{bmatrix} 0 & 1 \\ 0 & 0 \end{bmatrix}$

(b) $A = \begin{bmatrix} 0 & 1 \\ -1 & 0 \end{bmatrix}$

(c) $A = \begin{bmatrix} 1 & 0 & 0 \\ 0 & 1 & 0 \\ 0 & 1 & 2 \end{bmatrix}$

题 2.2　将以下系统的表达式变换为约当规范形

(a) $A = \begin{bmatrix} -2 & 1 \\ -17 & -4 \end{bmatrix}$, $b = \begin{bmatrix} 4 \\ 8 \end{bmatrix}$, $c = [0 \quad 1]$

(b) $A = \begin{bmatrix} 0 & 1 & 0 \\ 0 & 0 & 1 \\ 2 & 3 & 0 \end{bmatrix}$, $b = \begin{bmatrix} 0 \\ 0 \\ 9 \end{bmatrix}$, $c = [1 \quad 0 \quad 0]$

(c) $A = \begin{bmatrix} -2 & 2 & -1 \\ 0 & -2 & 0 \\ 1 & -4 & 0 \end{bmatrix}$, $b = \begin{bmatrix} 0 \\ 0 \\ 16 \end{bmatrix}$, $c = [1 \quad 0 \quad 0]$

(d) $A = \begin{bmatrix} 0 & 1 & 0 \\ 3 & 0 & 2 \\ -12 & -7 & -6 \end{bmatrix}$, $B = \begin{bmatrix} 2 & 3 \\ 1 & 5 \\ 7 & 1 \end{bmatrix}$, $C = \begin{bmatrix} 1 & 0 & 0 \\ 0 & 1 & 0 \end{bmatrix}$

题 2.3　(a) 确定矩阵的特征方程、特征值和特征向量

$$A = \begin{bmatrix} 0 & 1 & 0 & 0 \\ 0 & 0 & 1 & 0 \\ 0 & 0 & 0 & 1 \\ 1 & 0 & 0 & 0 \end{bmatrix}$$

(b) 对于(a) 中给定矩阵找出变换矩阵将其变换为约当形。

题 2.4　一个线性系统具有以下状态转移矩阵

$$\Phi(t) = \begin{bmatrix} 2e^{-t} - e^{-2t} & -2e^{-t} + 2e^{-2t} \\ -e^{-t} + e^{-2t} & -e^{-t} + 2e^{-2t} \end{bmatrix}$$

确定该系统的系统矩阵 $\boldsymbol{A}$。

题 2.5　检验以下矩阵是否满足作为状态转移矩阵的条件。如果有矩阵满足该条件，找出相应的系统矩阵 $\boldsymbol{A}$。

$$(\text{a})\ \boldsymbol{\Phi}(t)=\begin{bmatrix}2e^{-t}-e^{-2t} & -2e^{-t}+2e^{-2t}\\ e^{-t}-e^{-2t} & -e^{-t}+2e^{-2t}\end{bmatrix}$$

$$(\text{b})\ \boldsymbol{\Phi}(t)=\begin{bmatrix}1 & \dfrac{(1-2e^{-2t})}{2}\\ 0 & e^{-2t}\end{bmatrix}$$

$$(\text{c})\ \boldsymbol{\Phi}(t)=\begin{bmatrix}1 & 0 & 0\\ 0 & \sin t & \cos t\\ 0 & -\cos t & \sin t\end{bmatrix}$$

$$(\text{d})\ \boldsymbol{\Phi}(t)=\begin{bmatrix}e^{-t} & 0 & 0\\ 0 & e^{-5t} & e^{-5t}\\ 0 & e^{-t} & e^{-5t}\end{bmatrix}$$

题 2.6　一个二阶系统 $\dot{\boldsymbol{x}}=\boldsymbol{A}\boldsymbol{x}$ 对两个不同初始状态的响应分别为

$$\boldsymbol{x}(t)=\begin{bmatrix}e^{-3t}\\ -e^{-3t}\end{bmatrix}\ \text{when}\ \boldsymbol{x}(0)=\begin{bmatrix}1\\ -4\end{bmatrix}$$

$$\boldsymbol{x}(t)=\begin{bmatrix}e^{-2t}\\ -e^{-2t}\end{bmatrix}\ \text{when}\ \boldsymbol{x}(0)=\begin{bmatrix}2\\ -1\end{bmatrix}$$

确定系统矩阵 $\boldsymbol{A}$。

题 2.7　考虑状态方程

$$\dot{\boldsymbol{x}}=\begin{bmatrix}1 & 0\\ 1 & 1\end{bmatrix}\boldsymbol{x}+\begin{bmatrix}1\\ 1\end{bmatrix}u$$

用两种不同方法确定当输入 $u(t)=1(t)$ 时的状态向量 $\boldsymbol{x}(t)$，$t\geqslant 0$。假设初始状态为 $\boldsymbol{x}(0)=[1\quad 0]^{\mathrm{T}}$。

题 2.8　描述一个离散系统的差分方程为

$$y(k+2)+3y(k+1)+2y(k)=2u(k+1)+3u(k)$$

(a) 确定能控规范形状态空间表达式；

(b) 求输入为

$$u(k)=\begin{cases}1 & k=0,1\\ 0 & k\neq 0,1\end{cases}$$

时系统的响应。假设初始状态为零。

题 2.9　将以下连续系统离散化

$$\dot{\boldsymbol{x}}=\begin{bmatrix}0 & 1\\ 0 & 2\end{bmatrix}\boldsymbol{x}+\begin{bmatrix}0\\ 1\end{bmatrix}u,\ y=[1\quad 0]\boldsymbol{x}$$

假设采样周期为 $T=1$ s。

第 3 章　线性系统的能控性和能观测性

3.1　引　言

系统分析通常考虑两大问题：定量分析和定性分析。在定量分析中，我们感兴趣的问题是系统对于特定输入和初始条件的精确响应。在定性分析中，感兴趣的则是系统的一般性质。能控性和能观测性是线性控制系统的两个定性性质，它们在现代控制理论的理论研究和实际应用两个方面都起着重要的作用。能控性和能观测性的概念可以用一个例子予以粗略的说明。

例 3.1　考虑如图 3.1 所示桥式电路，其中输入为 v_i，输出为 $y = v_C$。该系统的状态变量选为电容器两端的电压 v_C 和通过电感器的电流 i_L。

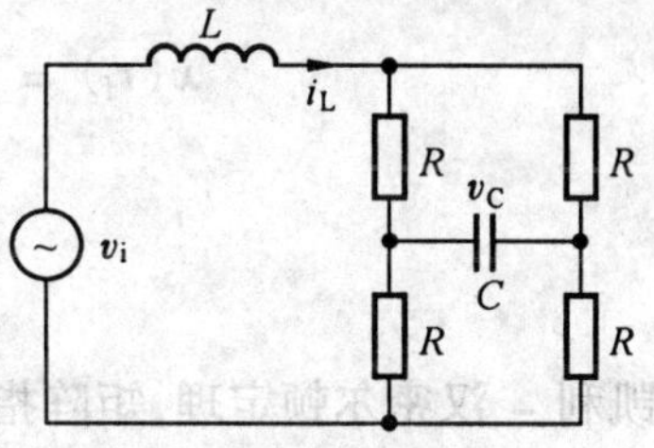

图 3.1　例 3.1 的桥式电路

令 $x_1 = i_L$、$x_2 = v_C$。由电路可知，如果 $x_2(t_0) = 0$，则无论施加什么输入，$t \geqslant t_0$ 时 $x_2(t) \equiv 0$，即输入对于 x_2 没有影响，此时称 x_2 为不能控的。如果输入为零，则无论 $x_1(t_0)$ 如何，$x_1(t)$ 的变化在输出 y 中不会有所反映，此时称 x_1 为不能观测的。

3.2　线性连续时间系统的能控性

3.2.1　能控性的定义

粗略地讲，能控性问题是研究由输入驱动状态的可能性。如果一个系统是能控的，由输入可以激励出该系统的所有响应模式。

考虑 n 维线性时不变连续系统

$$\left.\begin{aligned}\dot{\boldsymbol{x}}(t) &= \boldsymbol{A}\boldsymbol{x}(t) + \boldsymbol{B}\boldsymbol{u}(t)\\ \boldsymbol{y}(t) &= \boldsymbol{C}\boldsymbol{x}(t) + \boldsymbol{D}\boldsymbol{u}(t)\end{aligned}\right\} \tag{3.1}$$

其中 $\boldsymbol{A}, \boldsymbol{B}, \boldsymbol{C}$ 和 $\boldsymbol{D}$ 分别为 $n \times n, n \times p, q \times n$ 和 $q \times p$ 实常数矩阵。如果存在有限时间 $t_f > t_0$ 和时间区间 $[t_0, t_f]$ 上不受约束的控制 $\boldsymbol{u}(t)$，能使任一非零状态 $\boldsymbol{x}(t_0)$ 转移到 $\boldsymbol{x}(t_f) = \boldsymbol{0}$，则称该系统在时间 t_0 是状态完全能控的，或者简称是能控的。

能控性的这一定义仅仅要求控制 $\boldsymbol{u}(t)$ 能使初始状态转移到目标状态，对于应当遵循什么样的轨迹则并无规定。

3.2.2 能控性的秩判据

下面的状态能控性秩判据表明，能控性的条件与系统的系数矩阵 $\boldsymbol{A}$ 和 $\boldsymbol{B}$ 有关，即状态能控性仅仅是与状态方程有关的一种性质，输出方程在这里不起任何作用。该判据也给出了检验状态能控性的一种方法。

对于式(3.1) 中状态方程所描述的系统，状态完全能控的充分必要条件是 $n \times np$ 矩阵

$$\boldsymbol{Q}_{\mathrm{c}} = [\boldsymbol{B} \quad \boldsymbol{AB} \quad \cdots \quad \boldsymbol{A}^{n-1}\boldsymbol{B}] \tag{3.2}$$

的秩为 n。矩阵 $\boldsymbol{Q}_{\mathrm{c}}$ 由系数矩阵 $\boldsymbol{A}$ 和 $\boldsymbol{B}$ 表示，称为能控性矩阵。

证明 式(3.1) 中状态方程的解为

$$\boldsymbol{x}(t) = \mathrm{e}^{\boldsymbol{A}(t-t_0)}\boldsymbol{x}(t_0) + \int_{t_0}^{t}\mathrm{e}^{\boldsymbol{A}(t-\tau)}\boldsymbol{Bu}(\tau)\mathrm{d}\tau$$

不失一般性，可以假设 $t_0 = 0$。根据能控性的定义，如果系统能控则有

$$\boldsymbol{x}(t_{\mathrm{f}}) = \mathrm{e}^{\boldsymbol{A}t_{\mathrm{f}}}\boldsymbol{x}(0) + \int_{0}^{t_{\mathrm{f}}}\mathrm{e}^{\boldsymbol{A}(t_{\mathrm{f}}-\tau)}\boldsymbol{Bu}(\tau)\mathrm{d}\tau = 0$$

即

$$\boldsymbol{x}(0) = -\int_{0}^{t_{\mathrm{f}}}\mathrm{e}^{-\boldsymbol{A}\tau}\boldsymbol{Bu}(\tau)\mathrm{d}\tau \tag{3.3}$$

由凯利 - 汉密尔顿定理，矩阵指数函数可以写成

$$\mathrm{e}^{\boldsymbol{A}t} = \sum_{i=0}^{n-1}\alpha_i(t)\boldsymbol{A}^i \tag{3.4}$$

将式(3.4) 代入式(3.3) 并重新整理得到

$$\boldsymbol{x}(0) = -\int_{0}^{t_{\mathrm{f}}}\sum_{i=0}^{n-1}\alpha_i(\tau)\boldsymbol{A}^i\boldsymbol{Bu}(\tau)\mathrm{d}\tau = -\sum_{i=0}^{n-1}\boldsymbol{A}^i\boldsymbol{B}\int_{0}^{t_{\mathrm{f}}}\alpha_i(\tau)\boldsymbol{u}(\tau)\mathrm{d}\tau \tag{3.5}$$

令

$$\boldsymbol{f}_i = \int_{0}^{t_{\mathrm{f}}}\alpha_i(\tau)\boldsymbol{u}(\tau)\mathrm{d}\tau \tag{3.6}$$

则式(3.5) 变为

$$\boldsymbol{x}(0) = -\sum_{i=0}^{n-1}\boldsymbol{A}^i\boldsymbol{B}\boldsymbol{f}_i = -[\boldsymbol{B} \quad \boldsymbol{AB} \quad \cdots \quad \boldsymbol{A}^{n-1}\boldsymbol{B}]\begin{bmatrix}\boldsymbol{f}_0\\ \boldsymbol{f}_1\\ \vdots\\ \boldsymbol{f}_{n-1}\end{bmatrix} = -\boldsymbol{Q}_{\mathrm{c}}\boldsymbol{F}_{\mathrm{u}} \tag{3.7}$$

因此，能控性问题可以解释为：给定任意初始状态 $\boldsymbol{x}(t_0)$，寻找控制向量 $\boldsymbol{u}(t)$，对于有限时间 $t_{\mathrm{f}} > 0$ 使得最终状态 $\boldsymbol{x}(t_{\mathrm{f}}) = \boldsymbol{0}$。这个问题也就是对于给定的 $\boldsymbol{x}(t_0)$ 和矩阵 $\boldsymbol{Q}_{\mathrm{c}}$ 由式(3.7) 求解 $\boldsymbol{F}_{\mathrm{u}}$。因此，该系统状态完全能控的充分必要条件为矩阵 $\boldsymbol{Q}_{\mathrm{c}}$ 中存在一组 n 个线性无关列向量，或者说矩阵 $\boldsymbol{Q}_{\mathrm{c}}$ 的秩为 n。对于单输入系统，矩阵 $\boldsymbol{Q}_{\mathrm{c}}$ 是一个方阵，因此能控性的条件为 $\boldsymbol{Q}_{\mathrm{c}}$ 必须是非奇异的。

由于检验一个系统的状态能控性时只涉及矩阵 $\boldsymbol{A}$ 和 $\boldsymbol{B}$，有时说矩阵对$(\boldsymbol{A}, \boldsymbol{B})$ 能控

时,就意味着 $\boldsymbol{Q}_{\mathrm{c}}$ 的秩为 n,而且该系统是状态完全能控的。

例 3.2　给出系统的状态方程为

$$\dot{\boldsymbol{x}} = \begin{bmatrix} -1 & -2 & -2 \\ 0 & -1 & 1 \\ 1 & 0 & -1 \end{bmatrix} \boldsymbol{x} + \begin{bmatrix} 2 \\ 0 \\ 1 \end{bmatrix} u$$

检验该系统的状态能控性。

解　由给定的状态方程可得

$$\boldsymbol{AB} = \begin{bmatrix} -1 & -2 & -2 \\ 0 & -1 & 1 \\ 1 & 0 & -1 \end{bmatrix} \begin{bmatrix} 2 \\ 0 \\ 1 \end{bmatrix} = \begin{bmatrix} -4 \\ 1 \\ 1 \end{bmatrix}$$

$$\boldsymbol{A}^2\boldsymbol{B} = \begin{bmatrix} -1 & -2 & -2 \\ 0 & -1 & 1 \\ 1 & 0 & -1 \end{bmatrix} \begin{bmatrix} -4 \\ 1 \\ 1 \end{bmatrix} = \begin{bmatrix} 0 \\ 0 \\ -5 \end{bmatrix}$$

由此,能控性矩阵为

$$\boldsymbol{Q}_{\mathrm{c}} = [\boldsymbol{B} \quad \boldsymbol{AB} \quad \boldsymbol{A}^2\boldsymbol{B}] = \begin{bmatrix} 2 & -4 & 0 \\ 0 & 1 & 0 \\ 1 & 1 & -5 \end{bmatrix}$$

因为

$$\operatorname{rank}\boldsymbol{Q}_{\mathrm{c}} = 3 = n$$

该系统能控。

例 3.3　状态方程为

$$\dot{\boldsymbol{x}} = \begin{bmatrix} -2 & 0 \\ a & -3 \end{bmatrix} \boldsymbol{x} + \begin{bmatrix} 1 \\ 0 \end{bmatrix} u$$

确定能控性对于参数 a 的要求。

解　生成能控性矩阵得到

$$\boldsymbol{Q}_{\mathrm{c}} = \begin{bmatrix} 1 & -2 \\ 0 & a \end{bmatrix}$$

由于矩阵 $\boldsymbol{Q}_{\mathrm{c}}$ 的行列式等于 a,而且只有当 a 不为零时行列式才不为零,因此 $a \neq 0$ 时系统能控。

例 3.4　证明状态方程为能控规范形的单输入系统必定能控。

证明　记状态方程为

$$\dot{\boldsymbol{x}} = \begin{bmatrix} 0 & 1 & \cdots & 0 \\ \vdots & \vdots & & \vdots \\ 0 & 0 & \cdots & 1 \\ * & * & * & * \end{bmatrix} \boldsymbol{x} + \begin{bmatrix} 0 \\ \vdots \\ 0 \\ 1 \end{bmatrix} u$$

其中“*”表示可能非零的元。能控性矩阵为

$$Q_c=\begin{bmatrix}0&0&\cdots&1\\ \vdots&\vdots&&*\\ 0&1&\cdots&*\\ 1&*&\cdots&*\end{bmatrix}$$

显然,由于矩阵 Q_c 是一个三角阵,而三角阵一定是非奇异的,因此系统必定能控。

3.2.3 约当规范形的能控性判据

1.所有特征值互异

如果一个系统矩阵的所有特征值互异,则推导出的约当规范形状态方程为

$$\dot{x}=Ax+Bu=\begin{bmatrix}\lambda_1&&&\\&\lambda_2&&\\&&\ddots&\\&&&\lambda_n\end{bmatrix}x+Bu$$

在这种情况下,矩阵对(A,B)能控的充分必要条件是矩阵 B 中没有全零行。

例如,系统

$$\dot{x}=\begin{bmatrix}-2&0\\0&-3\end{bmatrix}x+\begin{bmatrix}1\\2\end{bmatrix}u$$

能控,而系统

$$\dot{x}=\begin{bmatrix}-2&0\\0&-3\end{bmatrix}x+\begin{bmatrix}0\\2\end{bmatrix}u$$

不能控。

2.特征值不全互异

假设推导出的约当形系数矩阵 A、B 和 C 由下式给出为

$$\left.\begin{aligned}
&\underset{(n\times n)}{A}=\begin{bmatrix}J_1&&&\\&J_2&&\\&&\ddots&\\&&&J_m\end{bmatrix},\quad \underset{(n\times p)}{B}=\begin{bmatrix}B_1\\B_2\\\vdots\\B_m\end{bmatrix},\quad \underset{(q\times n)}{C}=\begin{bmatrix}C_1&C_2&\cdots&C_m\end{bmatrix}\\
&\underset{(\sigma_i\times\sigma_i)}{J_i}=\begin{bmatrix}J_{i1}&&&\\&J_{i2}&&\\&&\ddots&\\&&&J_{i\alpha_i}\end{bmatrix},\quad \underset{(\sigma_i\times p)}{B_i}=\begin{bmatrix}B_{i1}\\B_{i2}\\\vdots\\B_{i\alpha_i}\end{bmatrix},\quad \underset{(q\times\sigma_i)}{C_i}=\begin{bmatrix}C_{i1}&C_{i2}&\cdots&C_{i\alpha_i}\end{bmatrix}\\
&\underset{(r_{ij}\times r_{ij})}{J_{ij}}=\begin{bmatrix}\lambda_i&1&&\\&\lambda_i&\ddots&\\&&\ddots&1\\&&&\lambda_i\end{bmatrix},\quad \underset{(r_{ij}\times p)}{B_{ij}}=\begin{bmatrix}b_{1ij}\\b_{2ij}\\\vdots\\b_{lij}\end{bmatrix},\quad \underset{(q\times r_{ij})}{C_{ij}}=\begin{bmatrix}c_{1ij}&c_{2ij}&\cdots&c_{lij}\end{bmatrix}
\end{aligned}\right\}\quad(3.8)$$

式中 $n\times n$ 矩阵A 为约当形,有 m 个互异特征值$\lambda_1,\lambda_2,\cdots,\lambda_m$,记号 J_i 表示所有与特征值

λ_i 相关的约当块，α_i 为 $\boldsymbol{J}_i$ 中约当块的块数，$\boldsymbol{J}_{ij}$ 为 $\boldsymbol{J}_i$ 中的第 j 约当块，且

$$n = \sum_{i=1}^{m} \sigma_i = \sum_{i=1}^{m} \sum_{j=1}^{\alpha_i} r_{ij}$$

$\boldsymbol{B}_{ij}$ 的第一行和最后一行分别记为 $\boldsymbol{b}_{1ij}$ 和 $\boldsymbol{b}_{lij}$。$\boldsymbol{C}_{ij}$ 的第一列和最后一列分别记为 $\boldsymbol{c}_{1ij}$ 和 $\boldsymbol{c}_{lij}$。

在这种情况下，系统能控的充分必要条件是 $i = 1,2,\cdots,m$ 时 $\alpha_i \times p$ 矩阵

$$\boldsymbol{B}_i^l = \begin{bmatrix} \boldsymbol{b}_{li1} \\ \boldsymbol{b}_{li2} \\ \vdots \\ \boldsymbol{b}_{li\alpha_i} \end{bmatrix} \tag{3.9}$$

的行是线性无关的。

例 3.5　判断以下系统的能控性

(1) $\dot{\boldsymbol{x}} = \begin{bmatrix} -2 & 0 \\ 0 & -2 \end{bmatrix} \boldsymbol{x} + \begin{bmatrix} 1 \\ 2 \end{bmatrix} u$；

(2) $\dot{\boldsymbol{x}} = \left[\begin{array}{ccc:cc} -2 & & & & \\ & -2 & 1 & & \\ & & -2 & & \\ \hdashline & & & -5 & 1 \\ & & & & -5 \end{array}\right] \boldsymbol{x} + \left[\begin{array}{cc} 1 & 0 \\ 0 & 2 \\ 0 & 1 \\ \hdashline 0 & 0 \\ 2 & 1 \end{array}\right] \boldsymbol{u}$；

(3) $\dot{\boldsymbol{x}} = \left[\begin{array}{cccc:ccc} -2 & 1 & & & & & \\ & -2 & & & & & \\ & & -2 & & & & \\ & & & -2 & & & \\ \hdashline & & & & 3 & 1 & \\ & & & & & 3 & \\ & & & & & & 3 \end{array}\right] \boldsymbol{x} + \left[\begin{array}{ccc} 0 & 0 & 0 \\ 1 & 0 & 0 \\ 0 & 3 & 0 \\ 0 & 0 & 2 \\ \hdashline 0 & 0 & 0 \\ 1 & 1 & 0 \\ 0 & 2 & 1 \end{array}\right] \boldsymbol{u}$。

解　(1) 只有一个互异特征值且 $\alpha_1 = -2$。由于

$$\boldsymbol{B}_1^l = \begin{bmatrix} 1 \\ 2 \end{bmatrix}$$

而这一矩阵中两个常数总是线性相关的，所以该系统不能控。

(2) 由于

$$\boldsymbol{B}_1^l = \begin{bmatrix} 0 & 2 \\ 0 & 1 \end{bmatrix}$$

式中两个行向量线性相关，因此该系统不能控。

(3) 由式(3.9) 有

$$\boldsymbol{B}_1^l = \begin{bmatrix} 1 & 0 & 0 \\ 0 & 3 & 0 \\ 0 & 0 & 2 \end{bmatrix}, \ \boldsymbol{B}_2^l = \begin{bmatrix} 1 & 1 & 0 \\ 0 & 2 & 1 \end{bmatrix}$$

由于 $\boldsymbol{B}_1^l$ 三个行向量线性无关，而且 $\boldsymbol{B}_2^l$ 的两个行向量也线性无关，因此该系统能控。

3.2.4 变换为能控规范形

如果一个单输入系统是能控的,那么该系统可以变换为能控规范形。

证明　考虑 n 维单输入系统

$$\dot{\boldsymbol{x}}(t) = \boldsymbol{A}\boldsymbol{x}(t) + \boldsymbol{b}u(t)$$

由于假设条件该系统是能控的,因此能控性矩阵 $\boldsymbol{Q}_c = [\boldsymbol{b}\quad \boldsymbol{A}\boldsymbol{b}\quad \cdots\quad \boldsymbol{A}^{n-1}\boldsymbol{b}]$ 满秩。假设矩阵 $\boldsymbol{A}$ 的特征多项式为

$$\Delta(\lambda) = |sI - A| = \lambda^n + a_1\lambda^{n-1} + \cdots + a_{n-1}\lambda + a_n \tag{3.10}$$

令 $\boldsymbol{x} = \boldsymbol{P}_c\boldsymbol{z}$,其中变换阵为

$$\boldsymbol{P}_c = [\boldsymbol{b}\quad \boldsymbol{A}\boldsymbol{b}\quad \cdots\quad \boldsymbol{A}^{n-1}\boldsymbol{b}]\begin{bmatrix} a_{n-1} & a_{n-2} & \cdots & a_1 & 1 \\ a_{n-2} & a_{n-3} & \cdots & 1 & 0 \\ \vdots & \vdots & & \vdots & \vdots \\ a_1 & 1 & \cdots & 0 & 0 \\ 1 & 0 & \cdots & 0 & 0 \end{bmatrix} \tag{3.11}$$

记

$$\boldsymbol{P}_c = [\boldsymbol{p}_{c1}\quad \boldsymbol{p}_{c2}\quad \cdots\quad \boldsymbol{p}_{cn}]$$

有

$$\begin{aligned} \boldsymbol{p}_{cn} &= \boldsymbol{b} \\ \boldsymbol{p}_{c,n-1} &= (\boldsymbol{A} + a_1\boldsymbol{I})\boldsymbol{b} \\ \boldsymbol{p}_{c,n-2} &= (\boldsymbol{A}^2 + a_1\boldsymbol{A} + a_2\boldsymbol{I})\boldsymbol{b} \\ &\vdots \\ \boldsymbol{p}_{c1} &= (\boldsymbol{A}^{n-1} + a_1\boldsymbol{A}^{n-2} + \cdots + a_{n-1}\boldsymbol{I})\boldsymbol{b} \end{aligned}$$

可以观察到

$$\begin{aligned} \boldsymbol{A}\boldsymbol{p}_{c1} = {} & \boldsymbol{A}a_{n-1}\boldsymbol{b} + a_{n-2}\boldsymbol{A}^2\boldsymbol{b} + \cdots + a_1\boldsymbol{A}^{n-1}\boldsymbol{b} + \boldsymbol{A}^n\boldsymbol{b} = \\ & (a_n\boldsymbol{I} + a_{n-1}\boldsymbol{A} + \cdots + a_1\boldsymbol{A}^{n-1} + \boldsymbol{A}^n)\boldsymbol{b} - a_n\boldsymbol{b} = -a_n\boldsymbol{b} = -a_n\boldsymbol{p}_{cn} = \\ & [\boldsymbol{p}_{c1}\quad \boldsymbol{p}_{c2}\quad \cdots\quad \boldsymbol{p}_{cn}][0\quad \cdots\quad 0\quad -a_n]^{\mathrm{T}} \\ \boldsymbol{A}\boldsymbol{p}_{c2} = {} & \boldsymbol{A}a_{n-2}\boldsymbol{b} + a_{n-3}\boldsymbol{A}^2\boldsymbol{b} + \cdots + a_1\boldsymbol{A}^{n-2}\boldsymbol{b} + \boldsymbol{A}^{n-1}\boldsymbol{b} = \\ & \boldsymbol{A}^{-1}(a_{n-2}\boldsymbol{A}^2 + a_{n-3}\boldsymbol{A}^3 + \cdots + a_1\boldsymbol{A}^{n-1} + \boldsymbol{A}^n)\boldsymbol{b} = \boldsymbol{A}^{-1}(-a_n\boldsymbol{I} - a_{n-1}\boldsymbol{A})\boldsymbol{b} = \\ & \boldsymbol{A}^{-1}(-a_{n-1}\boldsymbol{A} - a_n\boldsymbol{I})\boldsymbol{p}_{cn} = -a_{n-1}\boldsymbol{p}_{cn} - a_n\boldsymbol{A}^{-1}\boldsymbol{p}_{cn} = -a_{n-1}\boldsymbol{p}_{cn} + \boldsymbol{p}_{c1} = \\ & -a_n\boldsymbol{A}^{-1}\boldsymbol{p}_{cn} - a_{n-1}\boldsymbol{p}_{cn} = \boldsymbol{p}_{c1} - a_{n-1}\boldsymbol{p}_{cn} = \\ & [\boldsymbol{p}_{c1}\quad \boldsymbol{p}_{c2}\quad \cdots\quad \boldsymbol{p}_{cn}][1\quad 0\quad \cdots\quad 0\quad -a_{n-1}]^{\mathrm{T}} \\ \boldsymbol{A}\boldsymbol{p}_{ci} = {} & \boldsymbol{p}_{c,i-1} - a_{n-i+1}\boldsymbol{p}_{cn} \\ \boldsymbol{A}\boldsymbol{p}_{cn} = {} & \boldsymbol{p}_{c,n-1} - a_1\boldsymbol{p}_{cn} = [\boldsymbol{p}_{c1}\quad \boldsymbol{p}_{c2}\quad \cdots\quad \boldsymbol{p}_{cn}][0\quad \cdots\quad 0\quad 1\quad -a_n]^{\mathrm{T}} \end{aligned}$$

因此有

$$P_c^{-1}AP_c = P_c^{-1}[\,p_{c1}\quad p_{c2}\quad \cdots\quad p_{cn}\,]\begin{bmatrix} 0 & 1 & \cdots & 0 & 0 \\ 0 & 0 & \cdots & 0 & 0 \\ \vdots & \vdots & & \vdots & \vdots \\ 0 & 0 & \cdots & 0 & 1 \\ -a_n & -a_{n-1} & \cdots & -a_2 & -a_1 \end{bmatrix} =$$

$$\begin{bmatrix} 0 & 1 & \cdots & 0 & 0 \\ 0 & 0 & \cdots & 0 & 0 \\ \vdots & \vdots & & \vdots & \vdots \\ 0 & 0 & \cdots & 0 & 1 \\ -a_n & -a_{n-1} & \cdots & -a_2 & -a_1 \end{bmatrix}$$

$$P_c^{-1}B = P_c^{-1}p_{cn} = P_c^{-1}[\,p_{c1}\quad p_{c2}\quad \cdots\quad p_{cn}\,]\begin{bmatrix} 0 \\ \vdots \\ 0 \\ 1 \end{bmatrix} = \begin{bmatrix} 0 \\ \vdots \\ 0 \\ 1 \end{bmatrix}$$

例 3.6　将下面能控的单变量状态空间表达式变换为能控规范形。

$$\dot{x} = \begin{bmatrix} 1 & 0 \\ -1 & 2 \end{bmatrix}x + \begin{bmatrix} -1 \\ 1 \end{bmatrix}u$$

$$y = [0\quad 1]x$$

解　给定矩阵 A 的特征多项式为

$$\Delta(\lambda) = |\lambda I - A| = \lambda^2 - 3\lambda + 2$$

因此 $a_1 = -3$、$a_2 = 2$。能控性矩阵 Q_c 为

$$Q_c = \begin{bmatrix} -1 & -1 \\ 1 & 3 \end{bmatrix}$$

由式(3.10)和式(3.11)有

$$P_c = \begin{bmatrix} -1 & -1 \\ 1 & 3 \end{bmatrix}\begin{bmatrix} -3 & 1 \\ 1 & 0 \end{bmatrix} = \begin{bmatrix} 0.5 & 0.5 \\ 0 & 1 \end{bmatrix}$$

因此,引入变换 $x = P_c z$,结果得到的能控规范形为

$$\dot{z} = \begin{bmatrix} 0 & 1 \\ -2 & 3 \end{bmatrix}z + \begin{bmatrix} 0 \\ 1 \end{bmatrix}u$$

$$y = [0\quad 1]z$$

这一结论对于多输入系统仍然成立,即一个能控的状态空间表达式可以变换为能控规范形。但是与单输入系统不一样,多输入系统的能控规范形不是唯一的。

3.2.5　输出能控性

在实际的工程设计中,应当控制输出。与系统的状态能控性相似,可以定义一个系统输出向量的能控性。考虑 n 维线性时不变连续系统

$$\left.\begin{aligned} \dot{x}(t) &= Ax(t) + Bu(t) \\ y(t) &= Cx(t) + Du(t) \end{aligned}\right\} \tag{3.12}$$

其中 $\boldsymbol{A},\boldsymbol{B},\boldsymbol{C}$ 和 $\boldsymbol{D}$ 分别为 $n\times n, n\times p, q\times n$ 和 $q\times p$ 实常数矩阵。

如果存在有限时间 $t_f > t_0$ 和时间区间 $[t_0, t_f]$ 上不受限制的控制 $\boldsymbol{u}(t)$，能使系统的任一非零初始输出 $\boldsymbol{y}(t_0)$ 转移到 $\boldsymbol{y}(t_f) = \boldsymbol{0}$，则该系统称为是输出能控的。

可以证明，式(3.12) 所描述的系统输出能控的充分必要条件是 $q\times np$ 矩阵 $[\boldsymbol{CB}\quad \boldsymbol{CAB}\quad \cdots\quad \boldsymbol{CA}^{n-1}\boldsymbol{B}\quad \boldsymbol{D}]$ 满秩，即

$$\mathrm{rank}[\boldsymbol{CB}\quad \boldsymbol{CAB}\quad \cdots\quad \boldsymbol{CA}^{n-1}\boldsymbol{B}\quad \boldsymbol{D}] = q \tag{3.13}$$

式(3.13) 中的矩阵称为输出能控性矩阵。

例 3.7　如果系统的系数矩阵为

$$\boldsymbol{A} = \begin{bmatrix} -4 & 5 \\ 1 & 0 \end{bmatrix},\ \boldsymbol{B} = \begin{bmatrix} -5 \\ 1 \end{bmatrix},\ \boldsymbol{C} = [1\quad 1],\ \boldsymbol{D} = 1$$

确定系统的输出能控性和状态能控性。

解　计算输出能控性矩阵的秩得到

$$\mathrm{rank}[\boldsymbol{CB}\quad \boldsymbol{CAB}\quad \boldsymbol{D}] = \mathrm{rank}[-6\quad 30\quad 1] = 1$$

秩等于 q，因此该系统是输出能控的。然后计算能控性矩阵的秩得到

$$\mathrm{rank}[\boldsymbol{B}\quad \boldsymbol{AB}] = \mathrm{rank}\begin{bmatrix} -5 & 25 \\ 1 & -5 \end{bmatrix} = 1$$

显然，$\mathrm{rank}\,\boldsymbol{Q}_c < n = 2$，因此该系统不是(完全状态) 能控的。

状态能控性是对状态方程定义的，而输出能控性则是对输入 - 输出描述定义的。因此，这两个概念之间并没有必然联系。

3.3　线性连续系统的能观测性

3.3.1　能观测性的定义

粗略地讲，能观测性的问题是研究由输出估计状态的可能性。如果一个系统是能观测的，该系统所有的响应模式都能在输出中观测到。

对于一个由下式描述的 n 维线性时不变系统

$$\left.\begin{aligned} \dot{\boldsymbol{x}}(t) &= \boldsymbol{Ax}(t) + \boldsymbol{Bu}(t) \\ \boldsymbol{y}(t) &= \boldsymbol{Cx}(t) + \boldsymbol{Du}(t) \end{aligned}\right\} \tag{3.14}$$

如果存在有限时间 $t_1 > t_0$，系统的任一初始状态 $\boldsymbol{x}(t_0)$ 可以由时间区间 $[t_0, t_1]$ 上测量到的输出 $\boldsymbol{y}(t)$ 所确定，则该系统称为是(完全状态) 能观测的。不失一般性，可以假设 $t_0 = 0$。

由式(3.14) 有

$$\boldsymbol{x}(t) = \mathrm{e}^{\boldsymbol{A}t}\boldsymbol{x}(0) + \int_0^t \mathrm{e}^{\boldsymbol{A}(t-\tau)}\boldsymbol{Bu}(\tau)\mathrm{d}\tau$$

和

$$\boldsymbol{y}(t) = \boldsymbol{C}\mathrm{e}^{\boldsymbol{A}t}\boldsymbol{x}(0) + \boldsymbol{C}\int_0^t \mathrm{e}^{\boldsymbol{A}(t-\tau)}\boldsymbol{Bu}(\tau)\mathrm{d}\tau + \boldsymbol{Du}(t) \tag{3.15}$$

由于能观测性的概念是假设矩阵 $\boldsymbol{A},\boldsymbol{B},\boldsymbol{C},\boldsymbol{D}$ 和输入已知的情况下定义的，因此上述方程右侧的后两项是已知的，并且在观测性的研究中可以从整个输出里面分离出来。或者，为了方便起见通常在式(3.15)中选择 $\boldsymbol{u}(t)\equiv\boldsymbol{0}$。

3.3.2　能观测性的秩判据

现在考虑 n 维线性时不变连续系统

$$\left.\begin{aligned}\dot{\boldsymbol{x}}(t)&=\boldsymbol{A}\boldsymbol{x}(t)\\\boldsymbol{y}(t)&=\boldsymbol{C}\boldsymbol{x}(t)\end{aligned}\right\}\tag{3.16}$$

下面的判据给出了一种检验能观测性的方法。

式(3.16)所描述系统能观测的充分必要条件是矩阵

$$\boldsymbol{Q}_o=\begin{bmatrix}\boldsymbol{C}\\\boldsymbol{C}\boldsymbol{A}\\\vdots\\\boldsymbol{C}\boldsymbol{A}^{n-1}\end{bmatrix}\tag{3.17}$$

的秩为 n。矩阵 $\boldsymbol{Q}_o$ 由矩阵 $\boldsymbol{A}$ 和 $\boldsymbol{C}$ 描述，并称为能观测性矩阵。

证明　由式(3.15)，式(3.16)系统的输出为

$$\boldsymbol{y}(t)=\boldsymbol{C}\mathrm{e}^{\boldsymbol{A}t}\boldsymbol{x}(0)$$

应用式(3.4)得到

$$\boldsymbol{y}(t)=\boldsymbol{C}\sum_{i=0}^{n-1}\alpha_i(t)\boldsymbol{A}^i\boldsymbol{x}(0)=\sum_{i=0}^{n-1}\alpha_i(t)\boldsymbol{C}\boldsymbol{A}^i\boldsymbol{x}(0)$$

或

$$\boldsymbol{y}(t)=[\alpha_0\boldsymbol{I}\quad\alpha_1\boldsymbol{I}\quad\cdots\quad\alpha_{n-1}\boldsymbol{I}]\begin{bmatrix}\boldsymbol{C}\\\boldsymbol{C}\boldsymbol{A}\\\vdots\\\boldsymbol{C}\boldsymbol{A}^{n-1}\end{bmatrix}\tag{3.18}$$

因此，系统能观测的充分必要条件为在矩阵 $\boldsymbol{Q}_o$ 中存在一组 n 个线性无关行向量，即能观测性矩阵的秩是 n。对于一个单输出系统，矩阵 $\boldsymbol{Q}_o$ 是个方阵，因此能观测性的条件为矩阵 $\boldsymbol{Q}_o$ 必须是非奇异的。

由于检验系统的能观测性时只涉及矩阵 $\boldsymbol{A}$ 和 $\boldsymbol{C}$，有时候说矩阵对 $(\boldsymbol{A},\boldsymbol{C})$ 能观测，就意味着 $\boldsymbol{Q}_o$ 的秩是 n。

例 3.8　检验下式所描述系统的能观测性

$$\dot{\boldsymbol{x}}=\begin{bmatrix}0&1&0\\0&0&1\\-6&-11&-6\end{bmatrix}\boldsymbol{x}+\begin{bmatrix}0\\0\\1\end{bmatrix}u$$

$$y=[4\quad5\quad1]\boldsymbol{x}。$$

解　由给定的状态空间表达式可以得到

$$\boldsymbol{C}\boldsymbol{A}=[4\quad5\quad1]\begin{bmatrix}0&1&0\\0&0&1\\-6&-11&-6\end{bmatrix}=[-6\quad-7\quad-1]$$

$$CA^2 = [-6 \quad -7 \quad -1]\begin{bmatrix} 0 & 1 & 0 \\ 0 & 0 & 1 \\ -6 & -11 & -6 \end{bmatrix} = [6 \quad 5 \quad -1]$$

由此,能观测性矩阵为

$$Q_o = \begin{bmatrix} C \\ CA \\ CA^2 \end{bmatrix} = \begin{bmatrix} 4 & 5 & 1 \\ -6 & -7 & -1 \\ 6 & 5 & -1 \end{bmatrix}$$

由于

$$\mathrm{rank}\,Q_o = 2 < n = 3$$

该系统是不能观测的。

例 3.9　证明状态空间表达式为能观测规范形的单输出系统必定能观测。

证明　该系统的状态空间表达式可以表示成

$$\dot{x} = \begin{bmatrix} 0 & \cdots & 0 & * \\ 1 & \cdots & 0 & * \\ \vdots & & \vdots & \vdots \\ 0 & \cdots & 1 & * \end{bmatrix} x, \; y = [0 \quad \cdots \quad 0 \quad 1]$$

其中“ * ”代表可能非零的元素。能观测性矩阵为

$$Q_o = \begin{bmatrix} 0 & 0 & \cdots & 1 \\ \vdots & \vdots & & \vdots \\ 0 & 1 & \cdots & * \\ 1 & * & \cdots & * \end{bmatrix}$$

Q_o是一个三角阵,三角阵必定非奇异。因此,如果一个单输出系统是用能观测规范形描述的,该系统总是能观测的。

3.3.3　约当规范形能观测性判据

1.所有特征值互异

如果一个系统矩阵的所有特征值都是互异的,假设 $u = 0$ 时推导出的约当规范形为

$$\dot{x} = \begin{bmatrix} \lambda_1 & & & \\ & \lambda_2 & & \\ & & \ddots & \\ & & & \lambda_n \end{bmatrix} x$$

$$y = [c_1 \quad c_2 \quad \cdots \quad c_n] x$$

在这种情况下,矩阵对(A, C)能观测的充分必要条件为矩阵 C 中没有全零元的列。

例如,系统

$$\dot{x} = \begin{bmatrix} -2 & 0 \\ 0 & -3 \end{bmatrix} x$$

$$y = [1 \quad 2] x$$

是能观测的;而系统

$$\dot{x} = \begin{bmatrix} -2 & 0 \\ 0 & -3 \end{bmatrix} x$$
$$y = [1 \quad 0] x$$

则是不能观测的。

2.特征值不全互异

假设推导所得约当形的系数矩阵 A,B 和 C 如式(3.8)中所示。在这种情况下,系统能观测的充分必要条件是 $i = 1,2,\cdots,m$ 时 $q \times \alpha_i$ 矩阵

$$C_i^1 = [c_{1i1} \quad c_{1i2} \quad \cdots \quad c_{1i\alpha_i}] \tag{3.19}$$

的列是线性无关的。

例 3.10　判断以下系统的能观测性。

(1)$\dot{x} = \begin{bmatrix} -2 & 0 \\ 0 & -2 \end{bmatrix} x$;

$y = [1 \quad 2] x$

(2)$\dot{x} = \left[\begin{array}{ccc:cc} -2 & & & & \\ & -2 & 1 & & \\ & & -2 & & \\ \hdashline & & & -5 & 1 \\ & & & & -5 \end{array}\right] x$;

$y = \begin{bmatrix} 1 & 2 & 0 & 2 & 2 \\ 0 & 0 & 1 & 0 & 1 \end{bmatrix} x$

(3)$\dot{x} = \left[\begin{array}{cccc:ccc} 2 & 1 & & & & & \\ & 2 & & & & & \\ & & 2 & & & & \\ & & & 2 & & & \\ \hdashline & & & & 3 & 1 & \\ & & & & & 3 & \\ & & & & & & 3 \end{array}\right] x$

$y = \begin{bmatrix} 1 & 0 & 0 & 0 & 1 & 0 & 0 \\ 0 & 0 & 1 & 0 & 1 & 0 & 1 \\ 0 & 0 & 0 & 2 & 0 & 0 & 0 \end{bmatrix} x$。

解　(1) 这里只有一个互异特征值,$\alpha_1 = 2$。因为

$$C_1^1 = [1 \quad 2]$$

在这一矩阵中两个常数总是线性相关的,所以该系统是不能观测的。

(2) 因为

$$B_1^l = \begin{bmatrix} 1 & 2 \\ 0 & 0 \end{bmatrix}$$

其中两个列是线性相关的,因此该系统是不能观测的。

由式(3.19) 有

$$C_1^1=\begin{bmatrix}1&0&0\\0&1&0\\0&0&2\end{bmatrix},\ C_2^1=\begin{bmatrix}1&0\\1&1\\0&0\end{bmatrix}$$

因为 C_1^1 的三个列向量线性无关，C_2^1 的两个列也向量线性无关，因此该系统是能观测的。

3.3.4　变换为能观测规范形

与能控性的情况类似，可以证明，如果 n 维单输出系统

$$\dot{x}(t)=Ax(t)+BU(t)$$
$$y(t)=Cx(t)$$

是能观测的，则通过引入非奇异变换 $x=P_o z$ 可以使该系统变换为能观测规范形。变换矩阵 P_o 可根据下式得到

$$P_o^{-1}=\begin{bmatrix}a_{n-1}&a_{n-2}&\cdots&a_1&1\\a_{n-2}&a_{n-3}&\cdots&1&0\\\vdots&\vdots&&\vdots&\vdots\\a_1&1&\cdots&0&0\\1&0&\cdots&0&0\end{bmatrix}Q_o\tag{3.20}$$

其中 $a_1,a_2,\cdots,a_n$ 是矩阵A 的特征多项式的系数，而 Q_o 则是该系统的能观测性矩阵。

例 3.11　将下面能观测的单变量状态空间表达式变换为能观测规范形

$$\dot{x}=\begin{bmatrix}1&0\\-1&2\end{bmatrix}x+\begin{bmatrix}-1\\1\end{bmatrix}u$$
$$y=[0\quad 1]x$$

解　因为能观测性矩阵

$$Q_o=\begin{bmatrix}0&1\\-1&2\end{bmatrix}$$

满秩，该系统是能观测的。给定矩阵 A 的特征多项式为

$$\Delta(\lambda)=|\lambda I-A|=\lambda^2-3\lambda+2$$

令 $x=P_o z$，并选择

$$P_o^{-1}=\begin{bmatrix}-3&1\\1&0\end{bmatrix}\begin{bmatrix}0&1\\-1&2\end{bmatrix}=\begin{bmatrix}-1&-1\\0&1\end{bmatrix},\ P_o=\begin{bmatrix}-1&-1\\0&1\end{bmatrix}$$

结果所得能观测规范形为

$$\dot{z}=\begin{bmatrix}0&-2\\1&3\end{bmatrix}z+\begin{bmatrix}0\\1\end{bmatrix}u$$
$$y=[0\quad 1]x$$

与单输出系统不一样，多输出系统的能观测规范形不是唯一的。

3.4　离散时间系统的能控性和能观测性

在前几节中已经学习了连续时间系统的能控性和能观测性。对于离散时间系统可以得到非常相似的分析和结果，这里假设输入和输出只是在离散时刻上已知(或对其感兴趣)，这些离散时刻又都是某一基本单位 T 的整数倍，譬如说，

$$t_0 = 0, t_1 = T, t_2 = 2T, \cdots$$

还假设用 k 表示的时间参数取整数值，$k = 0, \pm 1, \pm 2, \cdots$

离散时间状态空间描述通常写成如下形式

$$\left.\begin{aligned} \boldsymbol{x}(k+1) &= \boldsymbol{A}\boldsymbol{x}(k) + \boldsymbol{B}\boldsymbol{u}(k) \\ \boldsymbol{y}(k) &= \boldsymbol{C}\boldsymbol{x}(k) \end{aligned}\right\} \tag{3.21}$$

3.4.1　线性离散时间系统的能控性

对于式(3.21)的离散时间系统，如果对于任一非零状态 $\boldsymbol{x}(l)$ 存在有限时刻 $m > l$，且在时间区间 $[l, m]$ 上存在相应的输入 $\boldsymbol{u}(k)$，使得 $\boldsymbol{x}(l)$ 转移到状态 $\boldsymbol{x}(m) = \boldsymbol{0}$，则说该离散时间系统在时刻 l 上是能控的。

如果假设系统的维数为 n，并定义一个能控性矩阵

$$\boldsymbol{Q}_{\mathrm{c}} = [\boldsymbol{B} \quad \boldsymbol{AB} \quad \cdots \quad \boldsymbol{A}^{n-1}\boldsymbol{B}] \tag{3.22}$$

则与连续系统的情况类似，可以获得离散时间系统能控性的秩判据。由式(3.21)所描述的系统能控的充分必要条件是

$$\operatorname{rank} \boldsymbol{Q}_{\mathrm{c}} = n \tag{3.23}$$

3.4.2　线性离散时间系统的能观测性

对于式(3.21)的离散时间系统，如果存在有限时刻 $m > l$ 使得任何非零状态 $\boldsymbol{x}(l)$ 可以由 $[l, m]$ 上的输出 $\boldsymbol{y}(k)$ 被唯一地确定，则说该离散时间系统是在时刻 l 上能观测的。

式(3.21)所描述的 n 维离散时间系统能观测的充分必要条件是能观测性矩阵的秩为 n，能观测性矩阵定义为

$$\boldsymbol{Q}_{\mathrm{o}} = \begin{bmatrix} \boldsymbol{C} \\ \boldsymbol{CA} \\ \vdots \\ \boldsymbol{CA}^{n-1} \end{bmatrix} \tag{3.24}$$

3.4.3　离散化系统的能控性和能观测性

下面的例子将表明一个离散化后系统的能控性和能观测性可能会改变。

例 3.12　设原来的连续时间系统为

$$\dot{\boldsymbol{x}} = \begin{bmatrix} 0 & \omega \\ -\omega & 0 \end{bmatrix} \boldsymbol{x} + \begin{bmatrix} 0 \\ 1 \end{bmatrix} u, \omega > 0$$

$$y = [1 \quad 0]\boldsymbol{x}$$

且采样周期为 T。确定离散化后系统的能控性和能观测性。

解 离散化前,因为

$$\text{rank}\ \boldsymbol{Q}_c = \text{rank}\begin{bmatrix} 0 & \omega \\ 1 & 0 \end{bmatrix} = 2 = n$$

$$\text{rank}\ \boldsymbol{Q}_o = \text{rank}\begin{bmatrix} 1 & 0 \\ 0 & \omega \end{bmatrix} = 2 = n$$

连续时间系统是能控且能观测的。

应用拉普拉斯变换可以得到矩阵指数函数为

$$\mathrm{e}^{\boldsymbol{A}t} = \mathscr{L}^{-1}\begin{bmatrix} s & -1 \\ \omega^2 & s \end{bmatrix}^{-1} = \mathscr{L}^{-1}\begin{bmatrix} \dfrac{s}{s^2+\omega^2} & \dfrac{1}{s^2+\omega^2} \\ \dfrac{-\omega^2}{s^2+\omega^2} & \dfrac{s}{s^2+\omega^2} \end{bmatrix} = \begin{bmatrix} \cos\omega t & \dfrac{1}{\omega}\sin\omega t \\ -\omega\sin\omega t & \cos\omega t \end{bmatrix}$$

由此,离散化系统的系统矩阵和输入矩阵为

$$\boldsymbol{A}^*(T) = \mathrm{e}^{\boldsymbol{A}T} = \begin{bmatrix} \cos\omega T & \dfrac{1}{\omega}\sin\omega T \\ -\omega\sin\omega T & \cos\omega T \end{bmatrix}$$

$$\boldsymbol{B}^*(T) = \int_0^T \mathrm{e}^{\boldsymbol{A}T}\boldsymbol{B}\,\mathrm{d}t = \int_0^T \begin{bmatrix} \cos\omega t & \dfrac{1}{\omega}\sin\omega t \\ -\omega\sin\omega t & \cos\omega t \end{bmatrix}\begin{bmatrix} 0 \\ 1 \end{bmatrix}\mathrm{d}t = \begin{bmatrix} \dfrac{1-\cos\omega T}{\omega^2} \\ \dfrac{\sin\omega t}{\omega} \end{bmatrix}$$

且离散化状态空间表达式为

$$\boldsymbol{x}(k+1) = \begin{bmatrix} \cos\omega T & \dfrac{1}{\omega}\sin\omega T \\ -\omega\sin\omega T & \cos\omega T \end{bmatrix}\boldsymbol{x}(k) + \begin{bmatrix} \dfrac{1-\cos\omega T}{\omega^2} \\ \dfrac{\sin\omega t}{\omega} \end{bmatrix}u(k)$$

$$y(k) = [1 \quad 0]\boldsymbol{x}(k)$$

能控性矩阵为

$$\boldsymbol{Q}_c = \begin{bmatrix} \dfrac{1-\cos\omega T}{\omega^2} & \dfrac{\cos\omega T - \cos^2\omega t + \sin^2\omega t}{\omega^2} \\ \dfrac{\sin\omega t}{\omega} & \dfrac{2\sin\omega T\cos\omega T - \sin\omega t}{\omega} \end{bmatrix} = \begin{bmatrix} 1-\cos\omega T & \cos\omega T - \cos 2\omega T \\ \sin\omega T & -\sin\omega T + \sin 2\omega T \end{bmatrix}$$

可以看到,离散化系统能控的充分必要条件是 $T \neq \dfrac{k\pi}{\omega}, k = 0, \pm 1, \pm 2, \cdots$

离散化系统的能观测性矩阵为

$$\boldsymbol{Q}_o = \begin{bmatrix} 1 & 0 \\ \cos\omega T & \dfrac{\sin\omega T}{\omega} \end{bmatrix}$$

显然,离散化系统能观测的充分必要条件是 $T \neq \dfrac{k\pi}{\omega}, k = 0, \pm 1, \pm 2, \cdots$

3.5　能控性、能观测性和传递函数

尽管能控性和能观测性是现代控制理论的概念，但它们与传递函数的性质有着密切的关系。

如果一个单变量系统的传递函数有零极点对消，根据状态变量选择的不同情况，该系统将或者不能控或者不能观测。如果一个单变量系统的传递函数没有零极点对消，该系统总是可以由状态空间表达式表示成能控和能观测的系统。

证明　一个特征值互异的单变量系统表示为

$$\left.\begin{aligned}\dot{\boldsymbol{x}}(t) &= \boldsymbol{A}\boldsymbol{x}(t) + \boldsymbol{b}u(t)\\ y &= \boldsymbol{c}\boldsymbol{x}(t)\end{aligned}\right\} \tag{3.25}$$

使矩阵 $\boldsymbol{A}$ 对角化，新的规范形的状态空间表达式为

$$\left.\begin{aligned}\dot{\boldsymbol{x}} &= \begin{bmatrix}\lambda_1 & & & \\ & \lambda_2 & & \\ & & \ddots & \\ & & & \lambda_n\end{bmatrix}\boldsymbol{x} + \begin{bmatrix}\beta_1\\ \beta_2\\ \vdots\\ \beta_n\end{bmatrix}u\\ y &= \begin{bmatrix}\gamma_1 & \gamma_2 & \cdots & \gamma_n\end{bmatrix}\boldsymbol{x}\end{aligned}\right\} \tag{3.26}$$

假设零初始条件并取拉普拉斯变换得到输出和输入之间的传递函数为

$$\frac{Y(s)}{U(s)} = \begin{bmatrix}\gamma_1 & \gamma_2 & \cdots & \gamma_n\end{bmatrix}\begin{bmatrix}s-\lambda_1 & & & \\ & s-\lambda_2 & & \\ & & \ddots & \\ & & & s-\lambda_n\end{bmatrix}^{-1}\begin{bmatrix}\beta_1\\ \beta_2\\ \vdots\\ \beta_n\end{bmatrix} =$$

$$\begin{bmatrix}\gamma_1 & \gamma_2 & \cdots & \gamma_n\end{bmatrix}\begin{bmatrix}\frac{1}{s-\lambda_1} & & & \\ & \frac{1}{s-\lambda_2} & & \\ & & \ddots & \\ & & & \frac{1}{s-\lambda_n}\end{bmatrix}\begin{bmatrix}\beta_1\\ \beta_2\\ \vdots\\ \beta_n\end{bmatrix} =$$

$$\frac{\gamma_1\beta_1}{s-\lambda_1} + \frac{\gamma_2\beta_2}{s-\lambda_2} + \cdots + \frac{\gamma_n\beta_n}{s-\lambda_n} = \sum_{i=1}^{n}\frac{\gamma_i\beta_i}{s-\lambda_i} \tag{3.27}$$

式(3.25)所描述系统能控和（或）能观测的充分必要条件是 $\beta_i \neq 0$ 和（或）$\gamma_i \neq 0, i = 1, 2, \cdots, n$。

由于矩阵 $\boldsymbol{A}$ 的所有特征值互异，传递函数可以改写成零极点形式为

$$\frac{Y(s)}{U(s)} = \frac{k(s-z_1)(s-z_2)\cdots(s-z_m)}{(s-\lambda_1)(s-\lambda_2)\cdots(s-\lambda_n)},\ n > m \tag{3.28}$$

该式部分分式展开为

$$\frac{Y(s)}{U(s)} = \sum_{i=1}^{n}\frac{\delta_i}{s-\lambda_i},\ i = 1,2,\cdots,n \tag{3.29}$$

其中 $\delta_i = \lim_{s \to \lambda_i}(s - \lambda_i)\dfrac{Y(s)}{U(s)}, i = 1,2,\cdots,n$。如果式(3.28) 中$\dfrac{Y(s)}{U(s)}$有一对或多对相同的零点和极点,例如 $\lambda_1 = z_1$ 时则式(3.29) 中有 $\delta_1 = 0$。

比较式(3.27) 和式(3.29) 得

$$\delta_i = \gamma_i \beta_i, i = 1,2,\cdots,n \tag{3.30}$$

因此,当 $\delta_i = 0$ 时 β_i 和(或)γ_i 为零,且状态 x_i 将不能控和(或)不能观测,即系统将不能控和(或)不能观测。当 $\delta_i \neq 0$ 时 β_i 和 γ_i 都不会为零,系统能控且能观测。

例 3.13 系统的传递函数为

$$\frac{Y(s)}{U(s)} = \frac{s + 2.5}{s^2 + 1.5s - 2.5}$$

考虑该系统的能控性和能观测性。

解 由于

$$\frac{Y(s)}{U(s)} = \frac{s + 2.5}{(s + 2.5)(s - 1)} = \frac{0}{s + 2.5} + \frac{1}{s - 1}$$

该传递函数在 $s = -2.5$ 处有一对相同的零点和极点。状态空间表达式可以写成三种规范形

(1) $\dot{\boldsymbol{x}} = \begin{bmatrix} 0 & 1 \\ 2.5 & -1.5 \end{bmatrix}\boldsymbol{x} + \begin{bmatrix} 0 \\ 1 \end{bmatrix}u, y = [1 \quad 0]\boldsymbol{x}$

(2) $\dot{\boldsymbol{x}} = \begin{bmatrix} 0 & 2.5 \\ 1 & -1.5 \end{bmatrix}\boldsymbol{x} + \begin{bmatrix} 1 \\ 0 \end{bmatrix}u, \ y = [0 \quad 1]\boldsymbol{x}$

(3) $\dot{\boldsymbol{x}} = \begin{bmatrix} -2.5 & 0 \\ 0 & 1 \end{bmatrix}\boldsymbol{x} + \begin{bmatrix} 0 \\ 1 \end{bmatrix}u, \ y = [0 \quad 1]\boldsymbol{x}$

在情况(1) 中,状态空间表达式为能控规范形,系统能控但不能观测,因为

$$\text{rank } \boldsymbol{Q}_\text{o} = \text{rank}\begin{bmatrix} 1 & 0 \\ 1 & 0 \end{bmatrix} = 1 < n$$

在情况(2) 中,状态空间表达式为能观测规范形,系统能观测但不能控,因为

$$\text{rank } \boldsymbol{Q}_\text{c} = \text{rank}\begin{bmatrix} 1 & 1 \\ 0 & 0 \end{bmatrix} = 1 < n$$

在情况(3) 中,状态空间表达式为约当规范形,很容易就可以得出结论,该系统既不能控也不能观测。

3.6 对偶原理

由前面的学习可知,能控性和能观测性在概念和形式上都很相似。这一现象是由系统的对偶特性所决定的。

3.6.1 对偶系统

两个线性系统$(\boldsymbol{A}_1, \boldsymbol{B}_1, \boldsymbol{C}_1)$和$(\boldsymbol{A}_2, \boldsymbol{B}_2, \boldsymbol{C}_2)$,如果 $\boldsymbol{A}_1^\text{T} = \boldsymbol{A}_2$、$\boldsymbol{B}_1^\text{T} = \boldsymbol{C}_2$、$\boldsymbol{C}_1^\text{T} = \boldsymbol{B}_2$,则称它们是对偶系统。这就是说,如果一个系统的描述是

$$\left.\begin{aligned}\dot{x}(t) &= Ax(t) + Bu(t)\\ y &= Cx(t)\end{aligned}\right\} \tag{3.31}$$

那么,它的对偶系统定义为

$$\left.\begin{aligned}\dot{z} &= A^{\mathrm{T}}z + C^{\mathrm{T}}v\\ w &= B^{\mathrm{T}}z\end{aligned}\right\} \tag{3.32}$$

其中 x 和 z 是 $n\times 1$ 状态向量;u 和 v 分别是 $p\times 1$ 和 $q\times 1$ 输入向量;y 和 w 分别是 $q\times 1$ 和 $p\times 1$ 输出向量;A 是 $n\times n$ 矩阵,B 是 $n\times p$ 矩阵,C 是 $q\times n$ 矩阵;而 A^{T},B^{T},C^{T} 分别是 A,B,C 的转置阵。

3.6.2　对偶原理

记 Q_{ci} 和 Q_{oi} 为第 i 个系统的能控性矩阵和能观测性矩阵,$i=1,2$。分别考虑对偶系统能控性矩阵的秩和能观测性矩阵的秩之间的关系,有

$$\begin{aligned}\operatorname{rank} Q_{o2} &= \operatorname{rank}[C_2^{\mathrm{T}}\quad (C_2A_2)^{\mathrm{T}}\quad \cdots\quad (C_2A_2^{n-1})^{\mathrm{T}}] =\\ &\operatorname{rank}[C_2^{\mathrm{T}}\quad A_2^{\mathrm{T}}C_2^{\mathrm{T}}\quad \cdots\quad (A_2^{n-1})^{\mathrm{T}}C_2^{\mathrm{T}}] =\\ &\operatorname{rank}[(B_1^{\mathrm{T}})^{\mathrm{T}}\quad (A_1^{\mathrm{T}})^{\mathrm{T}}(B_1^{\mathrm{T}})^{\mathrm{T}}\quad \cdots\quad ((A_1^{\mathrm{T}})^{n-1})^{\mathrm{T}}(B_1^{\mathrm{T}})^{\mathrm{T}}] =\\ &\operatorname{rank}[B_1\quad A_1B_1\quad \cdots\quad A_1^{n-1}B_1] =\\ &\operatorname{rank}Q_{c1}\end{aligned}$$

$$\begin{aligned}\operatorname{rank}Q_{c2} &= \operatorname{rank}[B_2\quad A_2B_2\quad \cdots\quad A_2^{n-1}B_2] =\\ &\operatorname{rank}[C_1^{\mathrm{T}}\quad A_1^{\mathrm{T}}C_1^{\mathrm{T}}\quad \cdots\quad (A_1^{\mathrm{T}})^{n-1}C_1^{\mathrm{T}}] =\\ &\operatorname{rank}[C_1^{\mathrm{T}}\quad (C_1A_1)^{\mathrm{T}}\quad \cdots\quad (C_1A_1^{n-1})^{\mathrm{T}}] =\\ &\operatorname{rank}Q_{o1}\end{aligned}$$

由上面的关系式不难得到以下的对偶原理:如果两个 n 维线性系统(A_1,B_1,C_1)和(A_2,B_2,C_2)是对偶系统,则

(1) 系统(A_1,B_1,C_1)能控的充分必要条件是系统(A_2,B_2,C_2)能观测;

(2) 系统(A_1,B_1,C_1)能观测的充分必要条件是系统(A_2,B_2,C_2)能控。

3.7　结构分解

如果系统(A,B,C)是不能控的,通常是有些状态变量能控而其他的状态变量不能控。与此类似,如果系统(A,B,C)是不能观测的,通常是有些状态变量能观测而其他的状态变量不能观测。从能控性和能观测性的角度来看,状态变量 x 可以分解成四组:既能控又能观测的 x_{co}、能控但不能观测的 $x_{c\bar{o}}$、不能控但能观测的 $x_{\bar{c}o}$、既不能控又不能观测的 $x_{\bar{c}\bar{o}}$,其中下标 c 用于表示能控,下标 $\bar{c}$ 表示不能控,下标 o 表示能观测,$\bar{o}$ 表示不能观测。相应地,系统也能分解成四个子系统,这称为结构分解。

对于约当规范形的状态空间表达式,系统可以通过观测进行结构分解。

例 3.14　将下面的状态空间表达式作结构分解

$$\dot{x} = \begin{bmatrix} \lambda_1 & 1 & 0 & 0 & 0 \\ 0 & \lambda_1 & 1 & 0 & 0 \\ 0 & 0 & \lambda_1 & 0 & 0 \\ 0 & 0 & 0 & \lambda_2 & 1 \\ 0 & 0 & 0 & 0 & \lambda_2 \end{bmatrix} \boldsymbol{x} + \begin{bmatrix} 0 \\ 1 \\ 0 \\ 0 \\ 1 \end{bmatrix} u$$

$$y = [0 \quad 1 \quad 1 \quad 0 \quad 1]\boldsymbol{x}$$

解 由矩阵对$(\boldsymbol{A},\boldsymbol{B})$可以看到状态变量$x_1,x_2,x_4$和$x_5$是能控的，$x_3$是不能控的。由矩阵对$(\boldsymbol{A},\boldsymbol{C})$可以看到状态变量$x_2,x_3$和$x_5$是能观测的，$x_1$和$x_4$是能控但不能观测的，$x_3$是不能控但能观测的。状态空间表达式可以改写为

$$\begin{bmatrix} \dot{x}_2 \\ \dot{x}_5 \\ \dot{x}_1 \\ \dot{x}_4 \\ \dot{x}_3 \end{bmatrix} = \begin{bmatrix} \lambda_1 & 0 & 0 & 0 & 1 \\ 0 & \lambda_1 & 0 & 0 & 0 \\ 1 & 0 & \lambda_1 & 0 & 0 \\ 0 & 1 & 0 & \lambda_2 & 0 \\ 0 & 0 & 0 & 0 & \lambda_2 \end{bmatrix} \begin{bmatrix} x_2 \\ x_5 \\ x_1 \\ x_4 \\ x_3 \end{bmatrix} + \begin{bmatrix} 1 \\ 1 \\ 0 \\ 0 \\ 0 \end{bmatrix} u$$

$$y = [1 \quad 1 \quad 0 \quad 0 \quad 1]\boldsymbol{x}$$

下面将不加证明地介绍规范分解方法。

3.7.1 按能控性分解

考虑n维线性时不变系统$(\boldsymbol{A},\boldsymbol{B},\boldsymbol{C})$，如果能控性矩阵的秩为$n_c(n_c < n)$，则存在非奇异变换$\boldsymbol{x} = \boldsymbol{T}_c\boldsymbol{z}$使得

$$\left.\begin{aligned} \begin{bmatrix} \dot{z}_c \\ \dot{z}_{\bar{c}} \end{bmatrix} &= \boldsymbol{T}_c^{-1}\boldsymbol{A}\boldsymbol{T}_c\begin{bmatrix} z_c \\ z_{\bar{c}} \end{bmatrix} + \boldsymbol{T}_c^{-1}\boldsymbol{B}\boldsymbol{u} = \begin{bmatrix} \bar{\boldsymbol{A}}_c & \bar{\boldsymbol{A}}_{12} \\ \boldsymbol{0} & \bar{\boldsymbol{A}}_{\bar{c}} \end{bmatrix}\begin{bmatrix} z_c \\ z_{\bar{c}} \end{bmatrix} + \begin{bmatrix} \bar{\boldsymbol{B}}_c \\ 0 \end{bmatrix}\boldsymbol{u} \\ \begin{bmatrix} \boldsymbol{y}_c \\ \boldsymbol{y}_{\bar{c}} \end{bmatrix} &= \boldsymbol{C}\boldsymbol{T}_c\begin{bmatrix} z_c \\ z_{\bar{c}} \end{bmatrix} = [\bar{\boldsymbol{C}}_c \quad \bar{\boldsymbol{C}}_{\bar{c}}]\begin{bmatrix} z_c \\ z_{\bar{c}} \end{bmatrix} \end{aligned}\right\} \tag{3.33}$$

其中$\boldsymbol{T}_c$是一个非奇异变换阵，它由$\boldsymbol{Q}_c$的n_c个线性无关列向量和$n - n_c$个完全任意选择的列向量构成。

式(3.33)表明，$z_{\bar{c}}$中的状态变量没有直接或间接地通过z_c中的状态变量受到输入$\boldsymbol{u}$的影响。因此状态向量$z_{\bar{c}}$是不能控的。结果所得能控子系统为

$$\left.\begin{aligned} \dot{z}_c &= \bar{\boldsymbol{A}}_c z_c + \bar{\boldsymbol{A}}_{12} z_{\bar{c}} + \bar{\boldsymbol{B}}_c\boldsymbol{u} \\ \boldsymbol{y}_c &= \bar{\boldsymbol{C}}_c z_c \end{aligned}\right\} \tag{3.34}$$

而不能控子系统则为

$$\left.\begin{aligned} \dot{z}_{\bar{c}} &= \bar{\boldsymbol{A}}_{\bar{c}} z_{\bar{c}} \\ \boldsymbol{y}_{\bar{c}} &= \bar{\boldsymbol{C}}_{\bar{c}} z_{\bar{c}} \end{aligned}\right\} \tag{3.35}$$

不仅如此，n_1维子系统$(\bar{\boldsymbol{A}}_c,\bar{\boldsymbol{B}}_c,\bar{\boldsymbol{C}}_c)$是能控的，而且具有和系统$(\boldsymbol{A},\boldsymbol{B},\boldsymbol{C})$相同的传递函数矩阵。

例 3.15 按能控性分解下面的状态空间表达式

$$\dot{\boldsymbol{x}} = \begin{bmatrix} 1 & 2 & -1 \\ 0 & 1 & 0 \\ 1 & -4 & 3 \end{bmatrix}\boldsymbol{x} + \begin{bmatrix} 0 \\ 0 \\ 1 \end{bmatrix}u, y = [1 \quad -1 \quad 1]\boldsymbol{x}。$$

解　能控性矩阵为

$$\boldsymbol{Q}_{\mathrm{c}} = \begin{bmatrix} 0 & -1 & -4 \\ 0 & 0 & 0 \\ 1 & 3 & 8 \end{bmatrix}$$

它的秩是 2。可以选择变换阵为

$$\boldsymbol{T}_{\mathrm{c}} = \begin{bmatrix} 0 & -1 & 0 \\ 0 & 0 & 0 \\ 1 & 3 & 1 \end{bmatrix}$$

其中前两列是 $\boldsymbol{Q}_{\mathrm{c}}$ 的线性无关列向量，而第三列则是任意选择的，只要使变换阵 $\boldsymbol{T}_{\mathrm{c}}$ 非奇异。$\boldsymbol{T}_{\mathrm{c}}$ 的逆阵为

$$\boldsymbol{T}_{\mathrm{c}}^{-1} = \begin{bmatrix} 3 & 0 & 1 \\ -1 & 0 & 0 \\ 0 & 1 & 0 \end{bmatrix}$$

令 $\boldsymbol{x} = \boldsymbol{T}_{\mathrm{c}}\boldsymbol{z}$ 得到

$$\dot{\boldsymbol{z}} = \begin{bmatrix} \dot{z}_{\mathrm{c}} \\ \dot{z}_{\bar{\mathrm{c}}} \end{bmatrix} = \begin{bmatrix} 0 & -4 & 2 \\ 1 & 4 & -2 \\ 0 & 0 & 1 \end{bmatrix}\begin{bmatrix} z_{\mathrm{c}} \\ z_{\bar{\mathrm{c}}} \end{bmatrix} + \begin{bmatrix} 1 \\ 0 \\ 0 \end{bmatrix}u$$

$$y = [1 \quad 2 \,\vdots\, -1]\begin{bmatrix} z_{\mathrm{c}} \\ z_{\bar{\mathrm{c}}} \end{bmatrix}$$

3.7.2　按能观测性分解

如果 n 维线性时不变系统 $(\boldsymbol{A},\boldsymbol{B},\boldsymbol{C})$ 的能观测性矩阵 $\boldsymbol{Q}_{\mathrm{o}}$ 的秩是 $n_{\mathrm{o}}(n_{\mathrm{o}} < n)$，则存在非奇异变换 $\boldsymbol{x} = \boldsymbol{T}_{\mathrm{o}}\boldsymbol{z}$ 使得

$$\left.\begin{aligned} \begin{bmatrix} \dot{z}_{\mathrm{o}} \\ \dot{z}_{\bar{\mathrm{o}}} \end{bmatrix} &= \boldsymbol{T}_{\mathrm{o}}^{-1}\boldsymbol{A}\boldsymbol{T}_{\mathrm{o}}\begin{bmatrix} z_{\mathrm{o}} \\ z_{\bar{\mathrm{o}}} \end{bmatrix} + \boldsymbol{T}_{\mathrm{o}}^{-1}\boldsymbol{B}\boldsymbol{u} = \begin{bmatrix} \tilde{\boldsymbol{A}}_{\mathrm{o}} & \boldsymbol{0} \\ \tilde{\boldsymbol{A}}_{21} & \tilde{\boldsymbol{A}}_{\bar{\mathrm{o}}} \end{bmatrix}\begin{bmatrix} z_{\mathrm{o}} \\ z_{\bar{\mathrm{o}}} \end{bmatrix} + \begin{bmatrix} \tilde{\boldsymbol{B}}_{\mathrm{o}} \\ \tilde{\boldsymbol{B}}_{\bar{\mathrm{o}}} \end{bmatrix}\boldsymbol{u} \\ \begin{bmatrix} \boldsymbol{y}_{\mathrm{o}} \\ \boldsymbol{y}_{\bar{\mathrm{o}}} \end{bmatrix} &= \boldsymbol{C}\boldsymbol{T}_{\mathrm{o}}\begin{bmatrix} z_{\mathrm{o}} \\ z_{\bar{\mathrm{o}}} \end{bmatrix} = [\tilde{\boldsymbol{C}}_{\mathrm{o}} \quad 0]\begin{bmatrix} z_{\mathrm{o}} \\ z_{\bar{\mathrm{o}}} \end{bmatrix} \end{aligned}\right\} \tag{3.36}$$

其中 $\boldsymbol{T}_{\mathrm{o}}^{-1}$ 是一个非奇异矩阵，而且可以由 $\boldsymbol{Q}_{\mathrm{o}}$ 的 n_{o} 个线性无关行和 $n - n_{\mathrm{o}}$ 个任意选择的行向量构成。

式(3.36) 表明，在 $z_{\bar{\mathrm{o}}}$ 中的状态变量没有直接或间接地通过 z_{o} 中的状态变量显现于输出 $\boldsymbol{y}$ 中。因此状态向量 $z_{\bar{\mathrm{o}}}$ 是不能观测的，最终所得的能观测子系统为

$$\left.\begin{aligned} \dot{z}_{\mathrm{o}} &= \tilde{\boldsymbol{A}}_{\mathrm{o}}z_{\mathrm{o}} + \tilde{\boldsymbol{B}}_{\mathrm{o}}\boldsymbol{u} \\ \boldsymbol{y}_{\mathrm{o}} &= \tilde{\boldsymbol{C}}_{\mathrm{o}}z_{\mathrm{o}} = \boldsymbol{y} \end{aligned}\right\} \tag{3.37}$$

而不能观测子系统则为

$$\left.\begin{aligned}\dot{z}_{\bar{o}} &= \tilde{A}_{21}z_o + \tilde{A}_{\bar{o}}z_{\bar{o}} + \tilde{B}_{\bar{o}}u \\ y_{\bar{o}} &= 0\end{aligned}\right\} \tag{3.38}$$

不仅如此，n_o 维子系统$(\tilde{A}_o,\tilde{B}_o,\tilde{C}_o)$是能观测的，而且具有和系统$(A,B,C)$一样的传递函数。

例 3.16 按能观测性分解由下面的状态空间表达式

$$\dot{x} = \begin{bmatrix}1 & 2 & -1\\0 & 1 & 0\\1 & -4 & 3\end{bmatrix}x + \begin{bmatrix}0\\0\\1\end{bmatrix}u,\quad y = \begin{bmatrix}1 & -1 & 1\end{bmatrix}x。$$

解 能观测性矩阵为

$$Q_o = \begin{bmatrix}1 & -1 & 1\\2 & -3 & 2\\4 & -7 & 4\end{bmatrix}$$

它的秩是 2。变换阵的逆阵可以选择为

$$T_o^{-1} = \begin{bmatrix}1 & -1 & 1\\2 & -3 & 2\\0 & 0 & 1\end{bmatrix}$$

其中前两行是 Q_o 的线性无关行，而第三行则是任意选择的，只要使得 T_o^{-1} 非奇异。变换阵为

$$T_o = \begin{bmatrix}3 & -1 & -1\\2 & -1 & 0\\0 & 0 & 1\end{bmatrix}$$

令 $x = T_o z$ 得到

$$\dot{z} = \begin{bmatrix}\dot{z}_o\\ \dot{z}_{\bar{o}}\end{bmatrix} = \left[\begin{array}{cc:c}0 & 1 & 0\\-2 & 3 & 0\\ \hdashline -5 & 3 & 2\end{array}\right]\begin{bmatrix}z_o\\ z_{\bar{o}}\end{bmatrix} + \left[\begin{array}{c}1\\2\\ \hdashline 1\end{array}\right]u$$

$$y = \left[\begin{array}{cc:c}1 & 0 & 0\end{array}\right]\begin{bmatrix}z_o\\ z_{\bar{o}}\end{bmatrix}$$

3.7.3 规范分解理论

考虑线性时不变系统

$$\begin{aligned}\dot{x}(t) &= Ax(t) + Bu(t)\\ y(t) &= Cx(t)\end{aligned}$$

通过非奇异变换系统(A,B,C)可以变换为以下规范形式

$$\left.\begin{aligned}\begin{bmatrix}\dot{w}_{co}\\ \dot{w}_{c\bar{o}}\\ \dot{w}_{\bar{c}o}\\ \dot{w}_{\bar{c}\bar{o}}\end{bmatrix} &= \begin{bmatrix}\hat{A}_{co} & 0 & \hat{A}_{13} & 0\\ \hat{A}_{21} & \hat{A}_{c\bar{o}} & \hat{A}_{23} & \hat{A}_{24}\\ 0 & 0 & \hat{A}_{\bar{c}o} & 0\\ 0 & 0 & \hat{A}_{43} & \hat{A}_{\bar{c}\bar{o}}\end{bmatrix}\begin{bmatrix}w_{co}\\ w_{c\bar{o}}\\ w_{\bar{c}o}\\ w_{\bar{c}\bar{o}}\end{bmatrix} + \begin{bmatrix}\hat{B}_{co}\\ \hat{B}_{c\bar{o}}\\ 0\\ 0\end{bmatrix}u\\ y &= [\hat{C}_{co} \quad 0 \quad \hat{C}_{\bar{c}o} \quad 0]\begin{bmatrix}w_{co}\\ w_{c\bar{o}}\\ w_{\bar{c}o}\\ w_{\bar{c}\bar{o}}\end{bmatrix}\end{aligned}\right\} \tag{3.39}$$

其中 w_{co} 是能控能观测的，$w_{c\bar{o}}$ 是能控但不能观测的，$w_{\bar{c}o}$ 是不能控但能观测的，$w_{\bar{c}\bar{o}}$ 是既不能控又不能观测的。

证明　如果系统(A,B,C)是不能控的，它可以变换为式(3.33)的形式。现在分别考虑系统(A,B,C)的能控和不能控部分。如果能控部分和(或)不能控部分是不能观测的，它们分别可以变换为式(3.36)的形式。把上述变换结合在一起，可以得到方程(3.39)。而且，系统(A,B,C)的传递函数为

$$G(s) = \hat{C}_{co}(sI - \hat{A}_{co})^{-1}\hat{B}_{co} \tag{3.40}$$

习　题

题 3.1　检查下列系统的能控性和能观测性。

(a) $A = \begin{bmatrix}1 & 0\\ -1 & 2\end{bmatrix}$, $b = \begin{bmatrix}1\\ 0\end{bmatrix}$, $c = [0 \quad 1]$

(b) $A = \begin{bmatrix}-2 & 2 & -1\\ 0 & -2 & 0\\ 1 & -4 & 0\end{bmatrix}$, $b = \begin{bmatrix}0\\ 0\\ 1\end{bmatrix}$, $c = [1 \quad -1 \quad 1]$

(c) $A = \begin{bmatrix}-3 & 1 & 0\\ 0 & -3 & 0\\ 0 & 0 & -1\end{bmatrix}$, $B = \begin{bmatrix}1 & -1\\ 0 & 0\\ 3 & 0\end{bmatrix}$, $C = \begin{bmatrix}1 & 0 & 0\\ 0 & 0 & 1\end{bmatrix}$

(d) $A = \begin{bmatrix}-5 & 0 & 0\\ 0 & -5 & 0\\ 0 & 0 & 1\end{bmatrix}$, $b = \begin{bmatrix}1\\ 3\\ 2\end{bmatrix}$, $C = \begin{bmatrix}1 & 2 & 3\\ 2 & 4 & 0\end{bmatrix}$

(e) $A = \begin{bmatrix}-2 & 0 & 0 & 0\\ 0 & -5 & 1 & 0\\ 0 & 0 & -5 & 1\\ 0 & 0 & 0 & -5\end{bmatrix}$, $b = \begin{bmatrix}2\\ 0\\ 0\\ 1\end{bmatrix}$, $c = [1 \quad 1 \quad 0 \quad 0]$

(f) $A = \begin{bmatrix}-2 & 0 & 0 & 0\\ 0 & -5 & 1 & 0\\ 0 & 0 & -5 & 1\\ 0 & 0 & 0 & -5\end{bmatrix}$, $b = \begin{bmatrix}1\\ 2\\ 0\\ 0\end{bmatrix}$, $c = [1 \quad 0 \quad 1 \quad 0]$

题 3.2 考虑系统

$$\dot{x}=\begin{bmatrix}a & 1\\0 & b\end{bmatrix}x+\begin{bmatrix}1\\1\end{bmatrix}u,\ y=[1\quad -1]x$$

寻求系统能控且能观测时参数 a 和 b 的取值范围。

题 3.3 确定以下系统是否输出能控。

(a) $A=\begin{bmatrix}0 & 1 & 0\\0 & 0 & 1\\-6 & -11 & -6\end{bmatrix},\ b=\begin{bmatrix}0\\0\\1\end{bmatrix},\ c=[1\quad 0\quad 0]$

(b) $A=\begin{bmatrix}-a & 0 & 0 & 0\\0 & -b & 0 & 0\\0 & 0 & -c & 0\\0 & 0 & 0 & -d\end{bmatrix},\ b=\begin{bmatrix}0\\0\\1\\1\end{bmatrix},\ c=[1\quad 0\quad 0\quad 0]$

题 3.4 系统的状态空间表达式为

$$\dot{x}=\begin{bmatrix}-1 & 0\\1 & -2\end{bmatrix}x+\begin{bmatrix}1\\-1\end{bmatrix}u$$

$$y=[-1\quad 1]x$$

(a) 将给定表达式通过适当的变换 $x=P_cz$ 变换为能控规范形。

(b) 将给定表达式通过适当的变换 $x=P_oz$ 变换为能观测规范形。

题 3.5 某控制系统的传递函数为

$$G(s)=\frac{s+a}{s^3+7s^2+14s+8}$$

确定使系统能控和(或)能观测时参数 a 的取值。

题 3.6 某控制系统的传递函数为

$$G(s)=\frac{s+1}{s^2+3s+2}$$

寻求状态空间表达式,使得系统是 (a) 能控但不能观测的;(b) 能观测但不能控的;(c) 既不能控也不能观测的。

题 3.7 考虑系统

$$\dot{x}=\begin{bmatrix}1 & 0 & 0 & 0\\0 & 2 & 0 & 0\\-6 & -2 & 3 & 0\\3 & -2 & 0 & 4\end{bmatrix}x+\begin{bmatrix}1\\0\\3\\1\end{bmatrix}u$$

$$y=[-4\quad -3\quad 1\quad 1]x$$

(a) 寻求能控子系统和不能控子系统。

(b) 寻求能观测子系统和不能观测子系统。

(c) 确定该系统的传递函数。

题 3.8 确定下列系统的能控性,将能控系统变换为能控规范形,将不能控系统变换为标准的不能控形式。

(a) $\boldsymbol{A} = \begin{bmatrix} -1 & 0 \\ 0 & -2 \end{bmatrix}$, $\boldsymbol{c} = [-1 \quad 1]$

(b) $\boldsymbol{A} = \begin{bmatrix} -2 & 2 \\ 0 & 24 \end{bmatrix}$, $\boldsymbol{c} = [1 \quad 1]$

(c) $\boldsymbol{A} = \begin{bmatrix} 1 & 2 & 0 \\ 3 & -1 & 1 \\ 0 & 2 & 0 \end{bmatrix}$, $\boldsymbol{b} = \begin{bmatrix} 1 \\ 0 \\ 1 \end{bmatrix}$

(d) $\boldsymbol{A} = \begin{bmatrix} 1 & 0 & 0 \\ 2 & 2 & 3 \\ -1 & 0 & 1 \end{bmatrix}$, $\boldsymbol{b} = \begin{bmatrix} 1 \\ 2 \\ -2 \end{bmatrix}$

题 3.9　确定下列系统的能观测性，将能观测系统变换为能观测规范形，将不能观测系统变换为标准的不能观测形式。

(a) $\boldsymbol{A} = \begin{bmatrix} 1 & 0 \\ -2 & 4 \end{bmatrix}$, $\boldsymbol{b} = \begin{bmatrix} 1 \\ 1 \end{bmatrix}$

(b) $\boldsymbol{A} = \begin{bmatrix} -1 & 0 \\ 1 & -2 \end{bmatrix}$, $\boldsymbol{b} = \begin{bmatrix} 1 \\ 1 \end{bmatrix}$

(c) $\boldsymbol{A} = \begin{bmatrix} 2 & 0 & 0 \\ 0 & 4 & 1 \\ 0 & 0 & 4 \end{bmatrix}$, $\boldsymbol{c} = [0 \quad 0 \quad 1]$

第4章　线性系统的状态综合

4.1　状态反馈和输出反馈

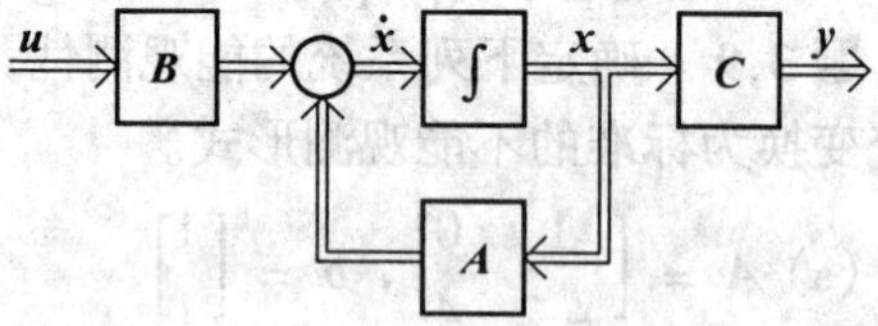

图4.1　开环系统的状态空间表达式

至今为止，在状态空间法的学习中只涉及如图4.1所示的开环系统，图中假设了直接传输矩阵 $\boldsymbol{D}=\boldsymbol{0}$。

考虑一个 n 维线性时不变连续系统

$$\left.\begin{aligned}\dot{\boldsymbol{x}}(t)&=\boldsymbol{A}\boldsymbol{x}(t)+\boldsymbol{B}\boldsymbol{u}(t)\\ \boldsymbol{y}(t)&=\boldsymbol{C}\boldsymbol{x}(t)\end{aligned}\right\}\tag{4.1}$$

其中 $\boldsymbol{x},\boldsymbol{u},\boldsymbol{y}$ 分别为 $n\times1$、$p\times1$、$q\times1$ 向量。由于现在系统的行为是用状态变量描述的，基于状态变量 $\boldsymbol{x}$ 和参考输入量 $\boldsymbol{v}$ 去选择控制量 $\boldsymbol{u}$ 是合理的，也就是说，一个好的控制应当由形式为 $\boldsymbol{u}(t)=\boldsymbol{f}(\boldsymbol{v}(t),\boldsymbol{x}(t))$ 的方程所确定。这一关系称为控制规律。

但是，对于设计控制系统而言，在经典控制理论和现代控制理论中反馈都是一种基本的方法。当一个系统用状态变量描述时，对于反馈控制系统而言有两种反馈：状态反馈和输出反馈。

4.1.1　状态反馈

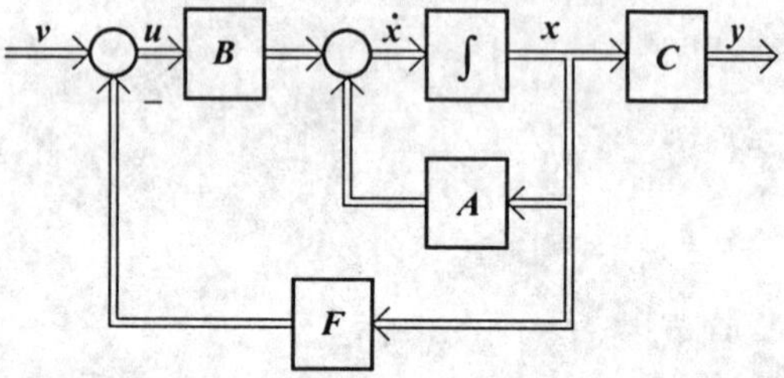

图4.2　状态反馈系统

如图4.2所示，在状态反馈中，每一个状态变量都乘以一个增益并反馈到输入端，图中 $\boldsymbol{F}$ 为反馈增益矩阵。

先研究一下引入状态反馈对系统状态空间表达式的影响。在图4.2中，控制规律由线性方程

$$\boldsymbol{u}=\boldsymbol{v}-\boldsymbol{F}\boldsymbol{x}\tag{4.2}$$

确定。将式(4.2)代入式(4.1)得到状态反馈系统的状态空间表达式为

$$\left.\begin{aligned}\dot{\boldsymbol{x}}(t)&=(\boldsymbol{A}-\boldsymbol{B}\boldsymbol{F})\boldsymbol{x}(t)+\boldsymbol{B}\boldsymbol{v}(t)\\ \boldsymbol{y}&=\boldsymbol{C}\boldsymbol{x}(t)\end{aligned}\right\}\tag{4.3}$$

而传递函数矩阵则由式

$$\boldsymbol{G}_{\mathrm{F}}(\mathrm{s})=\boldsymbol{C}(s\boldsymbol{I}-\boldsymbol{A}+\boldsymbol{B}\boldsymbol{F})^{-1}\boldsymbol{B}\tag{4.4}$$

给出。

4.1.2　输出反馈

在工程实践中，尤其是并非所有状态变量都可测量时，有时会采用输出反馈。典型的输出反馈如图 4.3 所示，其中 $\boldsymbol{H}$ 为反馈增益矩阵。

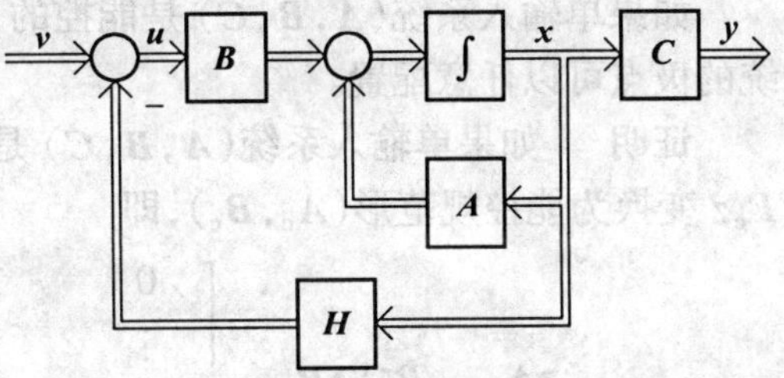

图 4.3　输出反馈系统

可以看到，现在的线性控制规律为

$$\boldsymbol{u} = \boldsymbol{v} - \boldsymbol{H}\boldsymbol{y} = \boldsymbol{v} - \boldsymbol{H}\boldsymbol{C}\boldsymbol{x} \tag{4.5}$$

因此，该输出反馈系统的状态空间表达式为

$$\left.\begin{aligned}\dot{\boldsymbol{x}} &= (\boldsymbol{A} - \boldsymbol{BHC})\boldsymbol{x} + \boldsymbol{B}\boldsymbol{v}\\ \boldsymbol{y} &= \boldsymbol{C}\boldsymbol{x}\end{aligned}\right\} \tag{4.6}$$

而传递函数矩阵则由式

$$\boldsymbol{G}_{\mathrm{H}}(s) = \boldsymbol{C}(s\boldsymbol{I} - \boldsymbol{A} + \boldsymbol{BHC})^{-1}\boldsymbol{B} \tag{4.7}$$

给出。

了解状态反馈和输出反馈的基本特点以及它们之间的差别是很有用的。

(1) 状态反馈和输出反馈两者都没有引入任何新的状态变量，即状态反馈和输出反馈没有改变它们原来开环系统的维数。

(2) 由于一般而言状态变量数大于输出变量数，通过输出反馈能得到的结果总是可以通过状态反馈得到，但是反过来不能这么讲。

(3) 在实际应用中，要检测到所有的状态并把它们反馈到输入端通常是很困难的。事实上，只有某些状态或者状态的组合能够作为输出进行测量。因此，任何实际的校正装置都一定是依赖于系统的输入、输出及一小部分可用于校正的状态变量。

4.1.3　状态反馈系统的能控性

状态反馈系统$(\boldsymbol{A} - \boldsymbol{BF}, \boldsymbol{B}, \boldsymbol{C})$能控的充分必要条件为系统$(\boldsymbol{A}, \boldsymbol{B}, \boldsymbol{C})$是能控的。

证明　我们先说明，如果$(\boldsymbol{A}, \boldsymbol{B}, \boldsymbol{C})$是能控的，那么$(\boldsymbol{A} - \boldsymbol{BF}, \boldsymbol{B}, \boldsymbol{C})$必定是能控的。令$\boldsymbol{x}(t_0)$是系统$(\boldsymbol{A}, \boldsymbol{B}, \boldsymbol{C})$的任意状态，且$x(t_f) = \boldsymbol{0}$。假设$(\boldsymbol{A}, \boldsymbol{B}, \boldsymbol{C})$是能控的，那么存在一个输入$\boldsymbol{u}(t)$，能在有限时间内使$\boldsymbol{x}(t_0)$转移到$\boldsymbol{x}(t_f)$。对于状态反馈系统$(\boldsymbol{A} - \boldsymbol{BF}, \boldsymbol{B}, \boldsymbol{C})$，如果现在选择$\boldsymbol{v} = \boldsymbol{u} + \boldsymbol{F}\boldsymbol{x}$，那么$\boldsymbol{v}(t)$将在有限时间内使$\boldsymbol{x}(t_0)$转移到$\boldsymbol{x}(t_f)$。因此，状态反馈系统$(\boldsymbol{A} - \boldsymbol{BF}, \boldsymbol{B}, \boldsymbol{C})$一定是能控的。

另一方面，由图 4.2 可以看到，控制状态的控制作用是由输入$\boldsymbol{v}(t)$产生的。因此，如果$\boldsymbol{u}(t)$不能控制$\boldsymbol{x}(t)$，那么$\boldsymbol{v}(t)$也不能控制$\boldsymbol{x}(t)$。也就是说，如果$(\boldsymbol{A}, \boldsymbol{B}, \boldsymbol{C})$是不能控的，那么$(\boldsymbol{A} - \boldsymbol{BF}, \boldsymbol{B}, \boldsymbol{C})$也是不能控的。

4.2　利用状态反馈配置极点

由于状态反馈和输出反馈都能改变闭环系统的系统矩阵，它们都能用来改变一个系统的极点位置。尤其是状态反馈可用于获取期望的极点位置。

4.2.1 极点配置的条件

如果单输入系统$(\boldsymbol{A},\boldsymbol{B},\boldsymbol{C})$是能控的,那么$(\boldsymbol{A}-\boldsymbol{BF},\boldsymbol{B},\boldsymbol{C})$的特征值,即状态反馈系统的极点可以任意配置。

证明 如果单输入系统$(\boldsymbol{A},\boldsymbol{B},\boldsymbol{C})$是能控的,那么它能够通过一个适当的变换$\boldsymbol{x}=\boldsymbol{P}_c\boldsymbol{z}$变换为能控规范形$(\boldsymbol{A}_c,\boldsymbol{B}_c)$,即

$$\boldsymbol{A}_c=\boldsymbol{P}_c^{-1}\boldsymbol{A}\boldsymbol{P}_c=\begin{bmatrix}0 & 1 & \cdots & 0\\ \vdots & \vdots & & \vdots\\ 0 & 0 & \cdots & 1\\ -\alpha_n & -\alpha_{n-1} & \cdots & -\alpha_1\end{bmatrix},\ \boldsymbol{B}_c=\boldsymbol{P}_c^{-1}\boldsymbol{B}=\begin{bmatrix}0\\ \vdots\\ 0\\ 1\end{bmatrix}$$

$$\boldsymbol{C}_c=\boldsymbol{C}\boldsymbol{P}_c=[\beta_n\ \ \beta_{n-1}\ \ \cdots\ \ \beta_1]$$

其中α_i和β_i,$i=1,2,\cdots,n$,是常数。通过观察可知,开环系统的传递函数为

$$\frac{Y(s)}{U(s)}=\boldsymbol{C}(s\boldsymbol{I}-\boldsymbol{A})^{-1}\boldsymbol{B}=\boldsymbol{C}_c(s\boldsymbol{I}-\boldsymbol{A}_c)^{-1}\boldsymbol{B}_c=\frac{\beta_1 s^{n-1}+\cdots+\beta_{n-1}s+\beta_n}{s^n+\alpha_1 s^{n-1}+\cdots+\alpha_{n-1}s+\alpha_n} \tag{4.8}$$

引入状态反馈

$$u=v-\boldsymbol{F}\boldsymbol{x}=v-\boldsymbol{F}\boldsymbol{P}_c\boldsymbol{z}=v-\boldsymbol{F}_z\boldsymbol{z} \tag{4.9}$$

其中

$$\boldsymbol{F}_z=\boldsymbol{F}\boldsymbol{P}_c=[\mathrm{f}_{z1}\ \ \mathrm{f}_{z2}\ \ \cdots\ \ \mathrm{f}_{zn}] \tag{4.10}$$

状态反馈系统的系统矩阵为

$$\boldsymbol{A}_c-\boldsymbol{B}_c\boldsymbol{F}_z=\begin{bmatrix}0 & 1 & \cdots & 0\\ \vdots & \vdots & & \vdots\\ 0 & 0 & \cdots & 1\\ -\alpha_n-f_{z1} & -\alpha_{n-1}-f_{z2} & \cdots & -\alpha_1-f_{zn}\end{bmatrix} \tag{4.11}$$

对于这一规范形,闭环系统的传递函数为

$$\frac{Y(s)}{V(s)}=\frac{\beta_1 s^{n-1}+\cdots+\beta_{n-1}s+\beta_n}{s^n+(\alpha_1+f_{zn})s^{n-1}+\cdots+(\alpha_{n-1}+f_{z2})s+(\alpha_n+f_{z1})} \tag{4.12}$$

由于特征多项式为

$$\Delta(s)=s^n+(\alpha_1+f_{zn})s^{n-1}+\cdots+(\alpha_{n-1}+f_{z2})s+(\alpha_n+f_{z1}) \tag{4.13}$$

可以得到结论,通过选择适当的状态反馈矩阵$\boldsymbol{F}_z$,矩阵$(\boldsymbol{A}_c-\boldsymbol{B}_c\boldsymbol{F}_z)$状态反馈系统的特征值、即闭环系统的极点可以任意配置。然后,由$\boldsymbol{F}_z$可以计算状态反馈矩阵$\boldsymbol{F}$。

4.2.2 极点配置的算法

1.直接比较特征多项式

假设系统$(\boldsymbol{A},\boldsymbol{B},\boldsymbol{C})$是能控的,状态反馈系统$(\boldsymbol{A}-\boldsymbol{BF},\boldsymbol{B},\boldsymbol{C})$的特征多项式为

$$\Delta(s)=|s\boldsymbol{I}-\boldsymbol{A}+\boldsymbol{BF}| \tag{4.14}$$

给定闭环系统的期望极点为$\hat{s}_1,\hat{s}_2,\cdots,\hat{s}_n$,那么所需的特征多项式应当为

$$\Delta_d(s) = (s - \hat{s}_1)(s - \hat{s}_2)\cdots(s - \hat{s}_n) = s^n + \hat{\alpha}_1 s^{n-1} + \cdots + \hat{\alpha}_{n-1}s + \hat{\alpha}_n \quad (4.15)$$

通过比较式(4.14)和式(4.15)的系数,可以得到状态反馈矩阵 $\boldsymbol{F}$ 的分量。

例 4.1　给定系统

$$\dot{\boldsymbol{x}} = \begin{bmatrix} 1 & 0 \\ -1 & 2 \end{bmatrix}\boldsymbol{x} + \begin{bmatrix} -1 \\ 1 \end{bmatrix}u$$

确定状态反馈矩阵 $\boldsymbol{F}$,使得闭环极点位于 $\hat{s}_{1,2} = -1 \pm j2$。

解　首先检验开环系统的能控性,

$$\text{rank } \boldsymbol{Q}_c = \text{rank}\begin{bmatrix} -1 & -1 \\ 1 & 3 \end{bmatrix} = 2$$

因此,该系统是能控的,而且状态反馈系统的极点能够任意配置。

记 $\boldsymbol{F} = [f_1 \quad f_2]$ 则

$$\boldsymbol{BF} = \begin{bmatrix} -1 \\ 1 \end{bmatrix}[f_1 \quad f_2] = \begin{bmatrix} -f_1 & -f_2 \\ f_1 & f_2 \end{bmatrix}$$

$$s\boldsymbol{I} - \boldsymbol{A} + \boldsymbol{BF} = \begin{bmatrix} s & 0 \\ 0 & s \end{bmatrix} - \begin{bmatrix} 1 & 0 \\ -1 & 2 \end{bmatrix} + \begin{bmatrix} -f_1 & -f_2 \\ f_1 & f_2 \end{bmatrix} = \begin{bmatrix} s-1-f_1 & -f_2 \\ 1+f_1 & s-2+f_2 \end{bmatrix}$$

取该矩阵的行列式得到特征多项式

$$|s\boldsymbol{I} - (\boldsymbol{A} - \boldsymbol{BF})| = s^2 + (-f_1 + f_2 - 3)s + (2f_1 + 2)$$

期望的特征多项式为

$$\Delta_d(s) = (s - \hat{s}_1)(s - \hat{s}_2) = s^2 + 2s + 5$$

使这两个式子的系数相等得到

$$-f_1 + f_2 - 3 = 2$$
$$2 + 2f_1 = 5$$

由此得到 $f_1 = 1.5$ 和 $f_2 = 6.5$,因此

$$\boldsymbol{F} = [1.5 \quad 6.5]$$

虽然这一方法对于阶数较低的系统相当简单,但是对于高于三阶的系统可能会变得很麻烦。对于这样的系统,另一种方法更好些。

2.用规范形

可以看到,4.2.1 节中给出的另一种方法是通过能控规范形确定状态反馈矩阵,使闭环极点放置在期望的位置上。由于矩阵 $\boldsymbol{F}_z$ 是关于状态 z 选择的,还必须用式(4.10)给出的 $\boldsymbol{F} = \boldsymbol{F}_z\boldsymbol{P}_c^{-1}$ 确定反馈矩阵 $\boldsymbol{F}$。

给定一个能控系统$(\boldsymbol{A}, \boldsymbol{B}, \boldsymbol{C})$和一组期望极点 $\hat{s}_1, \hat{s}_2, \cdots, \hat{s}_n$,确定状态反馈矩阵 $\boldsymbol{F}$ 的过程可以归纳如下:

(1) 求取矩阵 $\boldsymbol{A}$ 的特征多项式

$$\det(s\boldsymbol{I} - \boldsymbol{A}) = s^n + \alpha_1 s^{n-1} + \cdots + \alpha_{n-1}s + \alpha_n$$

(2) 计算期望特征多项式

$$\Delta_d(s) = (s - \hat{s}_1)(s - \hat{s}_2)\cdots(s - \hat{s}_n) = s^n + \hat{\alpha}_1 s^{n-1} + \cdots + \hat{\alpha}_{n-1}s + \hat{\alpha}_n$$

(3) 计算关于状态 z 的状态反馈矩阵

$$\boldsymbol{F}_z = [f_{z1} \quad f_{z2} \quad \cdots \quad f_{zn}] = [\hat{\alpha}_n - \alpha_n \quad \hat{\alpha}_{n-1} - \alpha_{n-1} \quad \cdots \quad \hat{\alpha}_1 - \alpha_1]$$

(4) 由式(3.11)计算将($\boldsymbol{A},\boldsymbol{B}$)变换为能控规范形的变换矩阵 $\boldsymbol{P}_c^{-1}$ 及其逆矩阵 $\boldsymbol{P}_c^{-1}$；

(5) 计算所需的状态反馈矩阵

$$\boldsymbol{F} = \boldsymbol{F}_z\boldsymbol{P}_c^{-1}。$$

例 4.2　仍然考虑例 4.1 中给出的系统

$$\dot{\boldsymbol{x}} = \begin{bmatrix} 1 & 0 \\ -1 & 2 \end{bmatrix}\boldsymbol{x} + \begin{bmatrix} -1 \\ 1 \end{bmatrix}u$$

用状态空间表达式的能控规范形求取状态反馈矩阵 $\boldsymbol{F}$，使得闭环极点位于 $\hat{s}_{1,2} = -1 \pm \mathrm{j}2$。

解　由例 4.1 可知，给定系统是能控的。

(1) 矩阵 $\boldsymbol{A}$ 的特征多项式为

$$\det(s\boldsymbol{I} - \boldsymbol{A}) = (s-1)(s-2) = s^2 - 3s + 2$$

(2) 期望的特征多项式为

$$\Delta_d(s) = (s - \hat{s}_1)(s - \hat{s}_2) = s^2 + 2s + 5$$

(3) 关于状态 z 的状态反馈矩阵为

$$\boldsymbol{F}_z = [f_{z1} \quad f_{z2}] = [\hat{\alpha}_2 - \alpha_2 \quad \hat{\alpha}_1 - \alpha_1] = [3 \quad 5]$$

(4) 变换矩阵 $\boldsymbol{P}_c$ 及其逆矩阵为

$$\boldsymbol{P}_c = \begin{bmatrix} -1 & -1 \\ 1 & 3 \end{bmatrix}\begin{bmatrix} -3 & 1 \\ 1 & 0 \end{bmatrix} = \begin{bmatrix} 2 & -1 \\ 0 & 1 \end{bmatrix},\ \boldsymbol{P}_c^{-1} = \begin{bmatrix} 0.5 & 0.5 \\ 0 & 1 \end{bmatrix}$$

(5) 所需的状态反馈矩阵为

$$\boldsymbol{F} = \boldsymbol{F}_z\boldsymbol{P}_c^{-1} = [3 \quad 5]\begin{bmatrix} 0.5 & 0.5 \\ 0 & 1 \end{bmatrix} = [1.5 \quad 6.5]$$

由于两个原因，这种状态反馈方法通常不太使用。一是状态反馈的结果得到 *PD* 或者 *PID* 校正装置，这些校正装置具有较宽的带宽；二是不可能或者难以检测到所有的状态并把它们反馈回去。事实上，常常只有某些状态或者它们的组合是可以测量并输出的。因此，任何实际使用的校正装置一定是只能依靠系统的输出量、输入量和少部分的状态变量。

4.3　镇定性

如果一个系统($\boldsymbol{A},\boldsymbol{B},\boldsymbol{C}$)是能控的，那么通过引入状态反馈，该系统的所有极点，即系统矩阵的所有特征值都可以任意配置。如果一个系统不是状态完全能控的，那么状态方程可以按能控性分解为以下形式

$$\begin{bmatrix} \dot{z}_c \\ \dot{z}_{\bar{c}} \end{bmatrix} = \begin{bmatrix} \overline{\boldsymbol{A}}_c & \overline{\boldsymbol{A}}_{12} \\ 0 & \overline{\boldsymbol{A}}_{\bar{c}} \end{bmatrix}\begin{bmatrix} z_c \\ z_{\bar{c}} \end{bmatrix} + \begin{bmatrix} \overline{\boldsymbol{B}}_c \\ 0 \end{bmatrix}\boldsymbol{u} \tag{4.16}$$

而且，降阶子系统

$$\dot{z}_c = \overline{\boldsymbol{A}}_c z_c + \overline{\boldsymbol{B}}_c\boldsymbol{u} \tag{4.17}$$

是能控的。由于系统矩阵形式的缘故，它的特征值是 $\overline{\boldsymbol{A}}_c$ 和 $\overline{\boldsymbol{A}}_{\bar{c}}$ 两者的特征值。考虑到输入

矩阵的形式,很容易看到 $\bar{A}_{\bar{c}}$ 的所有特征值都是不能控的。另一方面,通过引入状态反馈,$\bar{A}_c$ 的所有特征值都可以任意配置。

在设计一个系统时,有时候只需要将不稳定的特征值(具有非负实部的特征值)改变为稳定的特征值(具有负实部的特征值)。这称为镇定问题。由以上讨论可以看到,如果矩阵 $\bar{A}_{\bar{c}}$ 具有不稳定特征值,那么就不能使该系统镇定。

4.4 状态观测器

在前面的讨论中,状态反馈是在所有状态变量都能测量并可用做输出的假设条件下引入的。但是在实际应用中,由于技术上或者经济上的原因,这一假设条件并不总是成立。因此,为了系统能够应用状态反馈,常常需要为状态寻找到合理的替代。现在要讨论的问题是,只根据所知道的系统输入和输出去实际可行地获取该系统状态,并推导出一个真正的状态估计器,这种装置通常称为渐近观测器。

在状态观测器的讨论中,跟着变量的下标 e 用于表示是变量的估计量。例如,$\boldsymbol{x}_e$ 是 $\boldsymbol{x}$ 的估计量。

4.4.1 状态观测器

1.开环观测器

考虑线性时不变系统($\boldsymbol{A},\boldsymbol{B},\boldsymbol{C}$),假设状态变量是不可利用的。由于矩阵 $\boldsymbol{A},\boldsymbol{B},\boldsymbol{C}$ 是已知的,现在的问题是根据矩阵 $\boldsymbol{A},\boldsymbol{B},\boldsymbol{C}$ 提供的信息,由可以利用的输入和输出估计或者生成状态变量。一个简单的方法是利用集成电路或者计算机复制原来的系统,如图4.4所示。这一复制出来的系统称为开环观测器。

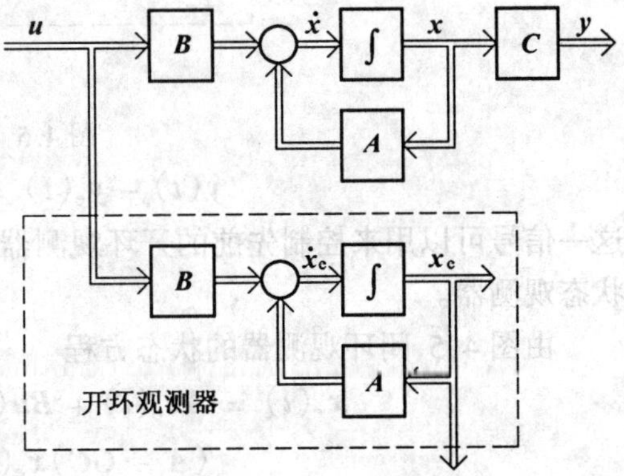

图 4.4　开环观测器

如果原来的系统和观测器具有相同的初始状态和相同的输入,那么观测器的输出 $\boldsymbol{x}_e(t)$ 对于所有时间都将等同于实际状态 $\boldsymbol{x}(t)$。注意,这里不再需要原来系统的输出。但是,不幸的是这一方法实际上存在很多缺点而难以对我们有所帮助。关键的问题是系统在运行期间会出现干扰,而这一方法则容易受到干扰的影响,而且对于任何误差都无法在事先确定的对策中加以补偿,这些误差有可能不太大但无法避免。

假设观测器的初始状态有着很小的误差

$$\boldsymbol{x}(t_0) - \boldsymbol{x}_e(t_0) = \varepsilon$$

由于原来的系统与观测器具有相同的输入,所以有

$$\dot{\boldsymbol{x}}(t) - \dot{\boldsymbol{x}}_e(t) = \boldsymbol{A}[\boldsymbol{x}(t) - \boldsymbol{x}_e(t)]$$

且

$$\boldsymbol{x}(t)-\boldsymbol{x}_e(t)=e^{\boldsymbol{A}t}[\boldsymbol{x}(t_0)-\boldsymbol{x}_e(t_0)]=e^{\boldsymbol{A}t}\cdot\epsilon$$

可以看到,如果系统是不稳定的,即矩阵 $\boldsymbol{A}$ 具有正实部特征值则无论初始误差多么小,实际状态 $\boldsymbol{x}(t)$ 和估计状态 $\boldsymbol{x}_e(t)$ 之间的误差将随时间而增大。即使系统是稳定的,如果矩阵 $\boldsymbol{A}$ 的特征值具有非常小的实部,初始误差的影响也将需要很长的时间才能消失。因此,通常情况下开环观测器是不能令人满意的。

2. 闭环观测器

要克服开环观测器可能存在的麻烦,典型的方法是利用反馈,即用与观测器误差成比例的信号操纵系统,使得这一误差为零。但是这里有一个问题是如何获取误差信号,因为 $\boldsymbol{x}(t)$ 是无法利用的。可以注意到,在开环观测器中没有用到系统的输出,而输出与状态密切相关又总是可以利用的。因此,可以生成误差信号为

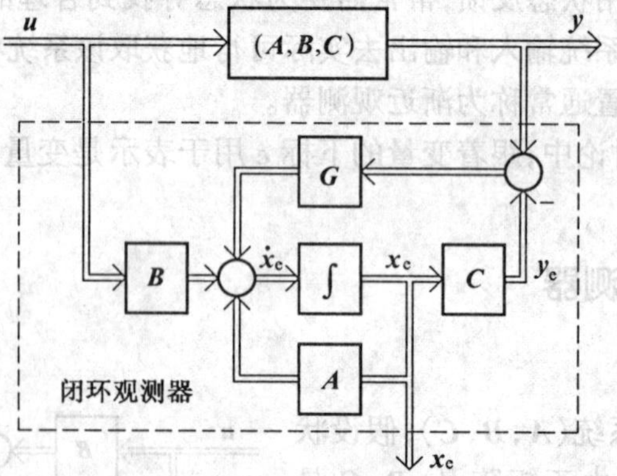

图 4.5　闭环观测器

$$\boldsymbol{y}(t)-\boldsymbol{y}_e(t)=\boldsymbol{C}[\boldsymbol{x}(t)-\boldsymbol{x}_e(t)]$$

这一信号可以用来控制先前的开环观测器。这一设想引导我们构成如图 4.5 所示的渐近状态观测器。

由图 4.5,闭环观测器的状态方程

$$\dot{\boldsymbol{x}}_e(t)=\boldsymbol{A}\boldsymbol{x}_e(t)+\boldsymbol{B}\boldsymbol{u}(t)+\boldsymbol{G}[\boldsymbol{y}(t)-\boldsymbol{C}\boldsymbol{x}_e(t)]=(\boldsymbol{A}-\boldsymbol{G}\boldsymbol{C})\boldsymbol{x}_e(t)+\boldsymbol{B}\boldsymbol{u}(t)+\boldsymbol{G}\boldsymbol{y}(t)\tag{4.18}$$

其中 $\boldsymbol{G}$ 是待适当选择的反馈增益矩阵。重新考虑 $\boldsymbol{x}(t)$ 和 $\boldsymbol{x}_e(t)$ 之间的误差有

$$\dot{\boldsymbol{x}}(t)-\dot{\boldsymbol{x}}_e(t)=\boldsymbol{A}\boldsymbol{x}(t)+\boldsymbol{B}\boldsymbol{u}(t)-[(\boldsymbol{A}-\boldsymbol{G}\boldsymbol{C})\boldsymbol{x}_e(t)+\boldsymbol{B}\boldsymbol{u}(t)+\boldsymbol{G}\boldsymbol{y}(t)]=(\boldsymbol{A}-\boldsymbol{G}\boldsymbol{C})[\boldsymbol{x}(t)-\boldsymbol{x}_e(t)]\tag{4.19}$$

以及

$$\boldsymbol{x}(t)-\boldsymbol{x}_e(t)=e^{(\boldsymbol{A}-\boldsymbol{G}\boldsymbol{C})t}[\boldsymbol{x}(t_0)-\boldsymbol{x}_e(t_0)]\tag{4.20}$$

由于 $\boldsymbol{y}(t)-\boldsymbol{y}_e(t)$ 的反馈提供有对于误差 $\boldsymbol{x}(t)-\boldsymbol{x}_e(t)$ 表现的控制,如果$(\boldsymbol{A}-\boldsymbol{G}\boldsymbol{C})$的所有特征根都具有负实部,那么无论观测器的初始状态 $\boldsymbol{x}_e(t_0)$ 是怎样的,误差 $\boldsymbol{x}(t)-\boldsymbol{x}_e(t)$ 将以渐近的方式趋近于零,即估计的状态 $\boldsymbol{x}_e(t)$ 将以渐近的方式趋近于实际状态 $\boldsymbol{x}(t)$。这就是为什么这种观测器称为渐近观测器的原因。简而言之,如果$(\boldsymbol{A}-\boldsymbol{G}\boldsymbol{C})$的特征值能够适当地加以选择,渐近状态观测器比开环观测器好得多。

4.4.2　状态观测器的设计

1. 配置观测器极点的条件

由式(4.20)可以看到，如果$(A - GC)$的所有特征值都具有负实部，则有

$$\lim_{t\to\infty}[x(t) - x_e(t)] = 0$$

而且其衰减速率取决于$(A - GC)$的特征值。问题在于是否总能找到一个合适的矩阵 G。答案为，如果(A, C)是能观测的则$(A - GC)$的特征值能够任意配置。

证明　根据对偶原理，如果(A, C)能观测，则(A^T, C^T)必定能控，并存在能使$(A^T - C^TF)$特征值任意配置的增益矩阵 F。由于转置不会改变一个矩阵的特征值，$(A - F^TC)$具有相同的特征值。因此，令 $G = F^T$，如果(A, C)能观测，则$(A - GC)$的特征值能够任意配置。

但是需要指出，应当考虑到特征值是否有一个大的负实部将关系到能否使系统不易受干扰影响。因此最佳的$(A - GC)$特征值，即观测器极点应当是观测器的响应速度和抗干扰能力之间的折中。

2. 观测器增益矩阵的算法

由于能控性和能观测性之间的对偶关系，观测器极点配置问题与我们讨论过的如何确定状态反馈矩阵 F 的问题基本上相似。

(1) 直接比较特征多项式

如果系统(A, B, C)是能观测的，由式(4.18)，观测器的状态方程为

$$\dot{x}_e(t) = (A - GC)x_e(t) + Bu(t) + Gy(t)$$

将观测器的特征多项式与期望特征多项式进行比较。观测器的特征多项式为

$$\Delta(s) = |sI - (A - GC)| \tag{4.21}$$

期望特征多项式为

$$\Delta_d(s) = (s - \hat{s}_1)(s - \hat{s}_2)\cdots(s - \hat{s}_n) = s^n + \hat{\alpha}_1 s^{n-1} + \cdots + \hat{\alpha}_{n-1}s + \hat{\alpha}_n \tag{4.22}$$

其中 $\hat{s}_1, \hat{s}_2, \cdots, \hat{s}_n$ 为闭环系统的期望极点。

例 4.3　给定系统

$$\dot{x} = \begin{bmatrix} 1 & 0 & 0 \\ 3 & -1 & 1 \\ 0 & 2 & 0 \end{bmatrix} x + \begin{bmatrix} 2 \\ 1 \\ 1 \end{bmatrix} u,$$

$$y = [0 \quad 0 \quad 1] x$$

设计一个渐近观测器，使其极点位于 $\hat{s}_1 = -3$、$\hat{s}_2 = -4$ 和 $\hat{s}_3 = -5$。

解　检查给定系统的能观测性矩阵有

$$\operatorname{rank} Q_o = \operatorname{rank} \begin{bmatrix} 0 & 0 & 1 \\ 0 & 2 & 0 \\ 6 & -2 & 2 \end{bmatrix} = 3$$

因此该系统是能观测的，观测器的极点可以任意配置。

记 $G = [g_1 \quad g_2 \quad g_2]^T$，得到

$$|sI - A + GC| = \left|\begin{bmatrix} s-1 & 0 & 0 \\ -3 & s+1 & -1 \\ 0 & -2 & s \end{bmatrix} + \begin{bmatrix} 0 & 0 & g_1 \\ 0 & 0 & g_2 \\ 0 & 0 & g_3 \end{bmatrix}\right| = \begin{vmatrix} s-1 & 0 & g_1 \\ -3 & s+1 & -1+g_2 \\ 0 & -2 & s+g_3 \end{vmatrix} =$$

$$s^3 + g_3 s^2 + (2g_2 - 3)s + (6g_1 - 2g_2 - g_3 + 2)$$

期望特征多项式为

$$\Delta_d(s) = (s+3)(s+4)(s+5) = s^3 + 12s^2 + 47s + 2$$

使这两个式子的系数相等，有

$$G = [g_1 \quad g_2 \quad g_3]^T = [20 \quad 25 \quad 12]^T$$

观测器的状态方程为

$$\dot{x}_e(t) = (A - GC)x_e(t) + Bu(t) + Gy(t) = \begin{bmatrix} 1 & 0 & -20 \\ 3 & -1 & -24 \\ 0 & 2 & -12 \end{bmatrix} x + \begin{bmatrix} 2 \\ 1 \\ 1 \end{bmatrix} u + \begin{bmatrix} 20 \\ 25 \\ 12 \end{bmatrix} y$$

(2) 用规范形

给定一个能观测的单输出系统

$$\left.\begin{aligned} \dot{x}(t) &= Ax(t) + Bu(t) \\ y(t) &= Cx(t) \end{aligned}\right\} \tag{4.23}$$

以及一组期望的观测器极点 $\hat{s}_1, \hat{s}_2, \cdots, \hat{s}_n$，确定反馈增益矩阵 G 的步骤归纳如下：

(1) 求取矩阵 A 的特征多项式

$$\det(sI - A) = s^n + \alpha_1 s^{n-1} + \cdots + \alpha_{n-1}s + \alpha_n \tag{4.24}$$

(2) 求取变换阵 P_o 和 P_o^{-1}，通过非奇异变换 $x = P_o z$ 使给定矩阵对 (A, C) 变换为规范形 (A_o, C_o)。由式(3.23)有

$$P_o^{-1} = \begin{bmatrix} a_{n-1} & a_{n-2} & \cdots & a_1 & 1 \\ a_{n-2} & a_{n-3} & \cdots & 1 & 0 \\ \vdots & \vdots & & \vdots & \vdots \\ a_1 & 1 & \cdots & 0 & 0 \\ 1 & 0 & \cdots & 0 & 0 \end{bmatrix} Q_o \tag{4.25}$$

(3) 计算期望特征多项式

$$\Delta_d(s) = (s - \hat{s}_1)(s - \hat{s}_2)\cdots(s - \hat{s}_n) = s^n + \hat{\alpha}_1 s^{n-1} + \cdots + \hat{\alpha}_{n-1}s + \hat{\alpha}_n \tag{4.26}$$

(4) 求取对于新的状态的反馈增益矩阵。记 $G_z = [g_{z1} \quad g_{z2} \quad \cdots \quad g_{zn}]^T$ 得到

$$|sI - (A_o - G_z C_o)| = \left| sI - \left[\begin{array}{c:c} \mathbf{0} & \begin{array}{c} -(\alpha_n + g_{z1}) \\ -(\alpha_{n-1} + g_{z2}) \end{array} \\ \hdashline I_{n-1} & \begin{array}{c} \vdots \\ -(\alpha_1 + g_{zn}) \end{array} \end{array}\right] \right| =$$

$$s^n + (\alpha_1 + g_{zn})s^{n-1} + \cdots + (\alpha_{n-1} + g_{z2})s + (\alpha_n + g_{z1})$$

将上面的多项式与式(4.26)比较得

$$G_z = [g_{z1} \quad g_{z2} \quad \cdots \quad g_{zn}]^T = [\hat{\alpha}_n - \alpha_n \quad \hat{\alpha}_{n-1} - \alpha_{n-1} \quad \cdots \quad \hat{\alpha}_1 - \alpha_1]^T \tag{4.27}$$

(5) 确定所需的反馈增益矩阵 G_o。引入变换 $x = P_o z$ 得到

$$\begin{aligned}\dot{z}_e &= P_o^{-1}(A - GC)P_o z_e + P_o^{-1}Bu + P_o^{-1}Gy = \\ &(P_o^{-1}AP_o - P_o^{-1}GCP_o)z_e + P_o^{-1}Bu + P_o^{-1}Gy = \\ &(A_o - G_z C_o)z_e + P_o^{-1}Bu + P_o^{-1}Gy\end{aligned}$$

因此,由上式得到,所需的反馈增益矩阵由下式给出

$$G = P_o G_z \tag{4.28}$$

例 4.4　仍然考虑给出的系统为

$$\dot{x} = \begin{bmatrix} 1 & 0 & 0 \\ 3 & -1 & 1 \\ 0 & 2 & 0 \end{bmatrix} x + \begin{bmatrix} 2 \\ 1 \\ 1 \end{bmatrix} u$$

$$y = \begin{bmatrix} 0 & 0 & 1 \end{bmatrix} x$$

用规范形的方法设计一个渐近观测器使其极点位于 $\hat{s}_1 = -3$、$\hat{s}_2 = -4$ 和 $\hat{s} = -5$。

解　由例 4.3 可知,该系统是能观测的。矩阵 A 的特征多项式为

$$\Delta(s) = \det(sI - A) = s^3 - 3s + 2$$

将(A, C) 变换为规范形(A_o, C_o) 的变换阵可求得为

$$P_o^{-1} = LQ_o = \begin{bmatrix} -3 & 0 & 1 \\ 0 & 1 & 0 \\ 1 & 0 & 0 \end{bmatrix} \begin{bmatrix} 0 & 0 & 1 \\ 0 & 2 & 0 \\ 6 & -2 & 2 \end{bmatrix} = \begin{bmatrix} 6 & -2 & -1 \\ 0 & 2 & 0 \\ 0 & 0 & 1 \end{bmatrix}$$

即

$$P_o = \frac{1}{12} \begin{bmatrix} 2 & 2 & 2 \\ 0 & 6 & 0 \\ 0 & 0 & 12 \end{bmatrix}$$

所需的特征多项式为

$$\Delta_d(s) = (s + 3)(s + 4)(s + 5) = s^3 + 12s^2 + 47s + 60$$

记 $G_z = [g_{z1} \quad g_{z2} \quad g_{z3}]^T$,有

$$G_z = [\hat{a}_3 - a_3 \quad \hat{a}_2 - a_2 \quad \hat{a}_1 - a_1]^T = [58 \quad 50 \quad 12]^T$$

因此,所需的反馈矩阵为

$$G = P_o G_z = \frac{1}{12} \begin{bmatrix} 2 & 2 & 2 \\ 0 & 6 & 0 \\ 0 & 0 & 12 \end{bmatrix} \begin{bmatrix} 58 \\ 50 \\ 12 \end{bmatrix} = \begin{bmatrix} 20 \\ 25 \\ 12 \end{bmatrix}$$

而观测器的状态方程则为

$$\dot{x}_e = (A - GC)x_e + Bu + Gy = \begin{bmatrix} 1 & 0 & -20 \\ 3 & -1 & -24 \\ 0 & 2 & -12 \end{bmatrix} x + \begin{bmatrix} 2 \\ 1 \\ 1 \end{bmatrix} u + \begin{bmatrix} 20 \\ 25 \\ 12 \end{bmatrix} y$$

3. 降维观测器

上面给出的观测器与被估计系统具有相同的维数。事实上,观测器的实际维数可以小于被估计系统的维数,因为输出已经给出了一些关于状态变量的信息。显然,设计一个维

数尽可能小的观测器是令人感兴趣的。下面将表明，如果输出矩阵的秩为 q，则可以设计一个维数为 $(n-q)$ 的渐近观测器用以生成所有的状态变量。这可以分两步完成。首先可以分离出 q 个直接由输出给出的状态变量。然后，借助于观测器估计出 $(n-q)$ 个状态变量。

考虑由下式描述的 n 维系统

$$\left.\begin{aligned}\dot{\boldsymbol{x}}(t) &= \boldsymbol{A}\boldsymbol{x}(t)+\boldsymbol{B}\boldsymbol{u}(t)\\ \boldsymbol{y}(t) &= \boldsymbol{C}\boldsymbol{x}(t)\end{aligned}\right\} \tag{4.29}$$

其中 $\boldsymbol{u}$ 和 $\boldsymbol{y}$ 分别是 $p\times 1$ 和 $q\times 1$ 向量。假设矩阵 $\boldsymbol{C}$ 的秩为 q。

构造一个 $n\times n$ 变换矩阵

$$\boldsymbol{P}=\begin{bmatrix}\boldsymbol{R}\\ \boldsymbol{C}\end{bmatrix}^{-1} \tag{4.30}$$

其中 $(n-q)\times n$ 实矩阵可以任意选择，只要使矩阵 $\boldsymbol{P}$ 是非奇异的。引入变换 $\boldsymbol{x}=\boldsymbol{P}\boldsymbol{z}$ 得到

$$\boldsymbol{CP}=\boldsymbol{C}\begin{bmatrix}\boldsymbol{R}\\ \boldsymbol{C}\end{bmatrix}^{-1}=[\boldsymbol{0}\quad \boldsymbol{I}]\begin{bmatrix}\boldsymbol{R}\\ \boldsymbol{C}\end{bmatrix}\begin{bmatrix}\boldsymbol{R}\\ \boldsymbol{C}\end{bmatrix}^{-1}=[\boldsymbol{0}\quad \boldsymbol{I}] \tag{4.31}$$

以及

$$\dot{\boldsymbol{z}}=\begin{bmatrix}\dot{\boldsymbol{z}}_1\\ \dot{\boldsymbol{z}}_2\end{bmatrix}=\begin{bmatrix}\bar{\boldsymbol{A}}_{11} & \bar{\boldsymbol{A}}_{12}\\ \bar{\boldsymbol{A}}_{21} & \bar{\boldsymbol{A}}_{22}\end{bmatrix}\begin{bmatrix}\boldsymbol{z}_1\\ \boldsymbol{z}_2\end{bmatrix}+\begin{bmatrix}\bar{\boldsymbol{B}}_1\\ \bar{\boldsymbol{B}}_2\end{bmatrix}\boldsymbol{u} \tag{4.32}$$

$$\boldsymbol{y}=[\boldsymbol{0}\quad \boldsymbol{I}]\begin{bmatrix}\boldsymbol{z}_1\\ \boldsymbol{z}_2\end{bmatrix}=\boldsymbol{z}_2 \tag{4.33}$$

式(4.33)表明 $(n-q)$ 状态变量 $\boldsymbol{z}_1$ 需要借助于一个观测器估计，因为 q 个状态变量 $\boldsymbol{z}_2$ 直接由输出给出。

将 $\boldsymbol{z}_2=\boldsymbol{y}$ 代入式(4.32)得到

$$\left.\begin{aligned}\dot{\boldsymbol{z}}_1 &= \bar{\boldsymbol{A}}_{11}\boldsymbol{z}_1+\bar{\boldsymbol{A}}_{12}\boldsymbol{z}_2+\bar{\boldsymbol{B}}_1\boldsymbol{u}=\bar{\boldsymbol{A}}_{11}\boldsymbol{z}_1+\bar{\boldsymbol{A}}_{12}\boldsymbol{y}+\bar{\boldsymbol{B}}_1\boldsymbol{u}\\ \dot{\boldsymbol{z}}_2 &= \bar{\boldsymbol{A}}_{21}\boldsymbol{z}_1+\bar{\boldsymbol{A}}_{22}\boldsymbol{z}_2+\bar{\boldsymbol{B}}_2\boldsymbol{u}=\bar{\boldsymbol{A}}_{21}\boldsymbol{z}_1+\bar{\boldsymbol{A}}_{22}\boldsymbol{y}+\bar{\boldsymbol{B}}_2\boldsymbol{u}=\dot{\boldsymbol{y}}\end{aligned}\right\} \tag{4.34}$$

对于与状态变量 $\boldsymbol{z}_1$ 有关的子系统，向量 $\boldsymbol{u}$ 和 $\boldsymbol{y}$ 可以看成是外部输入。这样，式(4.34)可以写成

$$\left.\begin{aligned}\dot{\boldsymbol{z}}_1 &= \bar{\boldsymbol{A}}_{11}\boldsymbol{z}_1+\bar{\boldsymbol{u}}\\ \bar{\boldsymbol{y}} &= \bar{\boldsymbol{A}}_{21}\boldsymbol{z}_1\end{aligned}\right\} \tag{4.35}$$

其中

$$\bar{\boldsymbol{u}}=\bar{\boldsymbol{A}}_{12}\boldsymbol{y}+\bar{\boldsymbol{B}}_1\boldsymbol{u} \tag{4.36}$$

$$\bar{\boldsymbol{y}}=\dot{\boldsymbol{y}}-\bar{\boldsymbol{A}}_{22}\boldsymbol{y}-\bar{\boldsymbol{B}}_2\boldsymbol{u} \tag{4.37}$$

如果系统 $(\boldsymbol{A},\boldsymbol{B},\boldsymbol{C})$ 是能观测的，则子系统 $(\bar{\boldsymbol{A}}_{11},\bar{\boldsymbol{A}}_{21})$ 必定是能观测的。这样，由式(4.18)可以得到一个 $(n-q)$ 维渐近观测器

$$\dot{\boldsymbol{z}}_1=(\bar{\boldsymbol{A}}_{11}-\boldsymbol{G}_1\bar{\boldsymbol{A}}_{21})\boldsymbol{z}_1+\bar{\boldsymbol{u}}+\boldsymbol{G}_1\bar{\boldsymbol{y}} \tag{4.38}$$

而且，通过选择一个合适的反馈增益矩阵 $\boldsymbol{G}_1$ 可以任意配置这一观测器的极点。

将式(4.36)和式(4.37)代入式(4.38)得到

$$\dot{\boldsymbol{z}}_{1e}=(\bar{\boldsymbol{A}}_{11}-\boldsymbol{G}_1\bar{\boldsymbol{A}}_{21})\boldsymbol{z}_{1e}+(\bar{\boldsymbol{A}}_{12}\boldsymbol{y}+\bar{\boldsymbol{B}}_1\boldsymbol{u})+\boldsymbol{G}_1(\dot{\boldsymbol{y}}-\bar{\boldsymbol{A}}_{22}\boldsymbol{y}-\bar{\boldsymbol{B}}_2\boldsymbol{u}) \tag{4.39}$$

可以看到,这一状态方程包含 $\dot{y}$。为了在构造观测器时避免使用微分器,引入一个新的状态向量

$$w_e = z_{1e} - G_1 y \tag{4.40}$$

由此,状态方程变为

$$\begin{aligned}\dot{w}_e = &(\bar{A}_{11} - G_1\bar{A}_{21})(w_e + G_1 y) + (\bar{A}_{12}y + \bar{B}_1 u) + G_1(\dot{y} - \bar{A}_{22}y - \bar{B}_2 u) - G_1\dot{y} = \\ &(\bar{A}_{11} - G_1\bar{A}_{21})w_e + (\bar{B}_1 - G_1\bar{B}_2)u + [(\bar{A}_{11} - G_1\bar{A}_{21})G_1 + \bar{A}_{12} - G_1\bar{A}_{22}]y\end{aligned} \tag{4.41}$$

最后,估计的状态向量如图 4.6 所示为

$$\dot{x}_e = Pz_e = P\begin{bmatrix} z_{1e} \\ y \end{bmatrix} = P\begin{bmatrix} w_e + G_1 y \\ y \end{bmatrix} \tag{4.42}$$

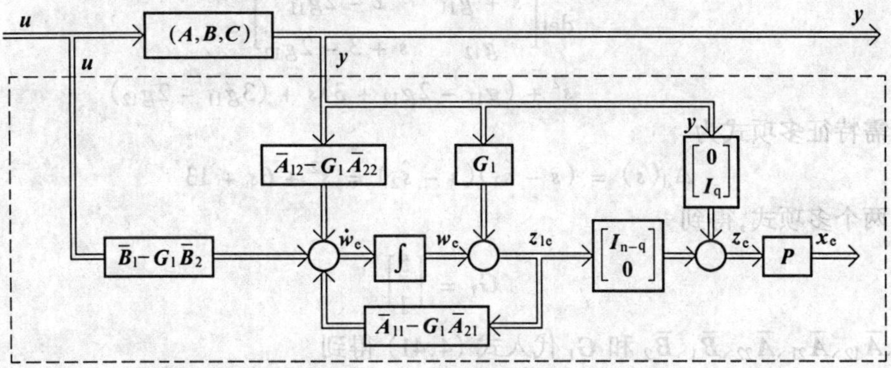

图 4.6　$(n - q)$ 维渐近状态观测器

例 4.5　考虑由下式给出的系统

$$\dot{x} = \begin{bmatrix} 0 & 0 & 1 \\ 0 & -1 & 1 \\ 1 & 2 & -1 \end{bmatrix} x + \begin{bmatrix} 2 \\ 1 \\ 1 \end{bmatrix} u$$

$$y = [0 \quad 2 \quad 1]x$$

设计一个降维观测器使其极点位于 $\hat{s}_{1,2} = -3 \pm \mathrm{j}2$。

解　检验该系统的能观测性有

$$\operatorname{rank} Q_o = \operatorname{rank}\begin{bmatrix} 0 & 2 & 1 \\ 1 & 0 & 1 \\ 1 & 2 & 0 \end{bmatrix} = 3$$

该系统是能观测的,而且可以构造一个二维的渐近状态观测器。

首先选择一个非奇异变换阵

$$P = \begin{bmatrix} R \\ C \end{bmatrix}^{-1} = \begin{bmatrix} 1 & 0 & 0 \\ 0 & 1 & 0 \\ 0 & 2 & 1 \end{bmatrix}^{-1} = \begin{bmatrix} 1 & 0 & 0 \\ 0 & 1 & 0 \\ 0 & -2 & 1 \end{bmatrix}$$

引入变换 $x = Pz$ 得到

$$\dot{z} = \begin{bmatrix} \bar{A}_{11} & \bar{A}_{12} \\ \bar{A}_{21} & \bar{A}_{22} \end{bmatrix} \begin{bmatrix} z_1 \\ z_2 \end{bmatrix} + \begin{bmatrix} \bar{B}_1 \\ \bar{B}_2 \end{bmatrix} u = P^{-1}APz + P^{-1}Bu =$$

$$\left[\begin{array}{cc:c} 0 & -2 & 1 \\ 0 & -3 & 1 \\ \hdashline 1 & -2 & 1 \end{array}\right] \begin{bmatrix} z_1 \\ z_2 \end{bmatrix} + \left[\begin{array}{c} 2 \\ 1 \\ \hdashline 3 \end{array}\right] u$$

$$y = CPz = [0 \quad 0 \,\vdots\, 1] z$$

由于 z_3 可以直接观测到,只需构造 z_1 和 z_2 的观测器。

由式(4.38),观测器的特征多项式为

$$\det(sI - (\bar{A}_{11} - G_1\bar{A}_{21})) = \det\left(\begin{bmatrix} s & \\ & s \end{bmatrix} - \begin{bmatrix} 0 & -2 \\ 0 & -3 \end{bmatrix} + \begin{bmatrix} g_{11} \\ g_{12} \end{bmatrix} [1 \quad -2] \right) =$$

$$\det\begin{bmatrix} s + g_{11} & 2 - 2g_{11} \\ g_{12} & s + 3 - 2g_{12} \end{bmatrix} =$$

$$s^2 + (g_{11} - 2g_{12} + 3)s + (3g_{11} - 2g_{12})$$

由于所需特征多项式为

$$\Delta_d(s) = (s - \hat{s}_1)(s - \hat{s}_2) = s^2 + 6s + 13$$

比较这两个多项式,得到

$$G_1 = \begin{bmatrix} 5 \\ 1 \end{bmatrix}$$

将 $\bar{A}_{11}$、$\bar{A}_{12}$、$\bar{A}_{21}$、$\bar{A}_{22}$、$\bar{B}_1$、$\bar{B}_2$ 和 G_1 代入式 (4.41) 得到

$$\dot{w}_e = (\bar{A}_{11} - G_1\bar{A}_{21})w_e + (\bar{B}_1 - G_1\bar{B}_2)u + [(\bar{A}_{11} - G_1\bar{A}_{21})G_1 + \bar{A}_{12} - G_1\bar{A}_{22}]y =$$

$$\left(\begin{bmatrix} 0 & -2 \\ 0 & -3 \end{bmatrix} - \begin{bmatrix} 5 \\ 1 \end{bmatrix} [1 \quad -2] \right) w_e + \left(\begin{bmatrix} 2 \\ 1 \end{bmatrix} - \begin{bmatrix} 5 \\ 1 \end{bmatrix} \cdot 3 \right) u +$$

$$\left\{ \left(\begin{bmatrix} 0 & -2 \\ 0 & -3 \end{bmatrix} - \begin{bmatrix} 5 \\ 1 \end{bmatrix} [1 \quad -2] \right) \begin{bmatrix} 5 \\ 1 \end{bmatrix} + \begin{bmatrix} 1 \\ 1 \end{bmatrix} - \begin{bmatrix} 5 \\ 1 \end{bmatrix} \cdot 1 \right\} y =$$

$$\begin{bmatrix} -5 & 8 \\ -1 & -1 \end{bmatrix} w_e + \begin{bmatrix} -13 \\ -2 \end{bmatrix} u + \begin{bmatrix} -21 \\ -6 \end{bmatrix} y$$

因此,由式 (4.40) 有

$$z_{1e} = \begin{bmatrix} z_{1e} \\ z_{2e} \end{bmatrix} = w_e + G_1 y = \begin{bmatrix} w_{1e} \\ w_{2e} \end{bmatrix} + \begin{bmatrix} 5 \\ 1 \end{bmatrix} y = \begin{bmatrix} w_{1e} + 5y \\ w_{2e} + y \end{bmatrix}$$

最后,估计的状态向量为

$$x_e = Pz_e = \begin{bmatrix} 1 & 0 & 0 \\ 0 & 1 & 0 \\ 0 & -2 & 1 \end{bmatrix} \begin{bmatrix} z_{1e} \\ z_{2e} \\ z_{3e} \end{bmatrix} = \begin{bmatrix} 1 & 0 & 0 \\ 0 & 1 & 0 \\ 0 & -2 & 1 \end{bmatrix} \begin{bmatrix} w_{1e} + 5y \\ w_{2e} + y \\ y \end{bmatrix} = \begin{bmatrix} w_{1e} + 5y \\ w_{2e} + y \\ -2w_{2e} - y \end{bmatrix}$$

4.4.3 分离原理

在状态反馈中使用估计状态 x_e 和实际状态 x 是否有什么差别,这是我们感兴趣的问题。

考虑一个能控且能观测的 n 维系统$(\boldsymbol{A},\boldsymbol{B},\boldsymbol{C})$,然后建立一个如图 4.7 所示组合系统,开始我们的讨论,图中补偿装置由一个全维观测器和一个状态反馈控制器组成。

这里有两个实际问题:

(1) 状态反馈矩阵是对于实际状态设计的,如果在反馈中使用估计状态能给出相同的结果吗?

(2) 在整个系统中呈现的状态观测器的特征值会有什么变化吗?

由图 4.7 通过观察可以写出以下方程

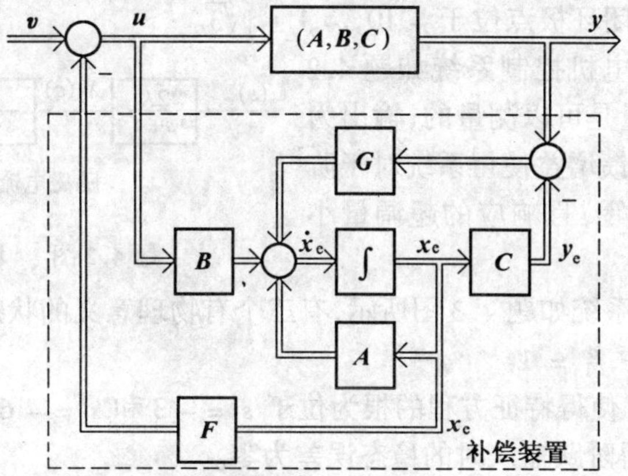

图 4.7　带有全维观测器的状态反馈系统

$$\dot{\boldsymbol{x}} = \boldsymbol{A}\boldsymbol{x} - \boldsymbol{B}\boldsymbol{F}\boldsymbol{x}_e + \boldsymbol{B}\boldsymbol{v}$$

$$\dot{\boldsymbol{x}}_e = \boldsymbol{G}\boldsymbol{C}\boldsymbol{x} + (\boldsymbol{A} - \boldsymbol{G}\boldsymbol{C} - \boldsymbol{B}\boldsymbol{F})\boldsymbol{x}_e + \boldsymbol{B}\boldsymbol{v}$$

而且,整个系统的状态方程可以安排成如下矩阵形式

$$\begin{bmatrix} \dot{\boldsymbol{x}} \\ \dot{\boldsymbol{x}}_e \end{bmatrix} = \begin{bmatrix} \boldsymbol{A} & -\boldsymbol{B}\boldsymbol{F} \\ \boldsymbol{G}\boldsymbol{C} & \boldsymbol{A} - \boldsymbol{G}\boldsymbol{C} - \boldsymbol{B}\boldsymbol{F} \end{bmatrix} \begin{bmatrix} \boldsymbol{x} \\ \boldsymbol{x}_e \end{bmatrix} + \begin{bmatrix} \boldsymbol{B} \\ \boldsymbol{B} \end{bmatrix} \boldsymbol{v} \tag{4.43}$$

引入下列非奇异变换

$$\begin{bmatrix} \boldsymbol{x} \\ \boldsymbol{x}_e \end{bmatrix} = \begin{bmatrix} \boldsymbol{I}_n & \boldsymbol{0} \\ \boldsymbol{I}_n & -\boldsymbol{I}_n \end{bmatrix} \begin{bmatrix} \boldsymbol{z} \\ \boldsymbol{z}_e \end{bmatrix}$$

式(4.43) 变为

$$\begin{bmatrix} \dot{\boldsymbol{z}} \\ \dot{\boldsymbol{z}}_e \end{bmatrix} = \begin{bmatrix} \boldsymbol{A} - \boldsymbol{B}\boldsymbol{F} & -\boldsymbol{B}\boldsymbol{F} \\ \boldsymbol{0} & \boldsymbol{A} - \boldsymbol{G}\boldsymbol{C} \end{bmatrix} \begin{bmatrix} \boldsymbol{z} \\ \boldsymbol{z}_e \end{bmatrix} + \begin{bmatrix} \boldsymbol{B} \\ \boldsymbol{0} \end{bmatrix} \boldsymbol{v} \tag{4.44}$$

由式(4.44) 马上可以看到,整个系统的闭环特征多项式为

$$\Delta(s) = |s\boldsymbol{I} - (\boldsymbol{A} - \boldsymbol{B}\boldsymbol{F})|\,|s\boldsymbol{I} - (\boldsymbol{A} - \boldsymbol{G}\boldsymbol{C})| \tag{4.45}$$

该式表明,在状态反馈中使用估计状态和实际状态,就我们所关心的特征值并没有什么差别。由此,状态反馈的设计和状态观测器的设计可以分别独立地进行。这一特性常常称为分离特性。

习　题

题 4.1　控制系统的状态方程为

$$\dot{x} = \begin{bmatrix} 0 & 1 & 0 \\ 0 & -1 & 1 \\ 0 & -1 & 10 \end{bmatrix} x + \begin{bmatrix} 0 \\ 0 \\ 10 \end{bmatrix} u$$

确定反馈增益矩阵使闭环极点位于 $-10, -1 \pm j\sqrt{3}$。

题 4.2　某交流电机控制系统如题 4.2 图所示。三个状态变量是可以测量的，输出为位置 $x_1(t)$。试选择反馈增益使得系统对于阶跃输入的稳态误差为零，且响应的超调量小于 3%。

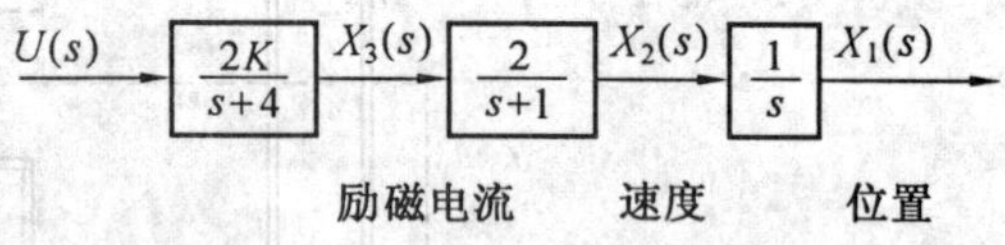

题 4.2 图　场控交流电机

题 4.3　某控制系统如题 4.3 图所示，有三个有物理意义的状态变量。状态反馈的结构已在图中画出，其中 $f_1 = 1$。

(a) 选择 f_2 和 f_3 使得特征方程的根为位于 $s = -3$ 和 $s = -6$ 之间的三个实根。

(b) 选择 K_P 使得阶跃输入时的稳态误差为零。

(c) 画出状态变量图。

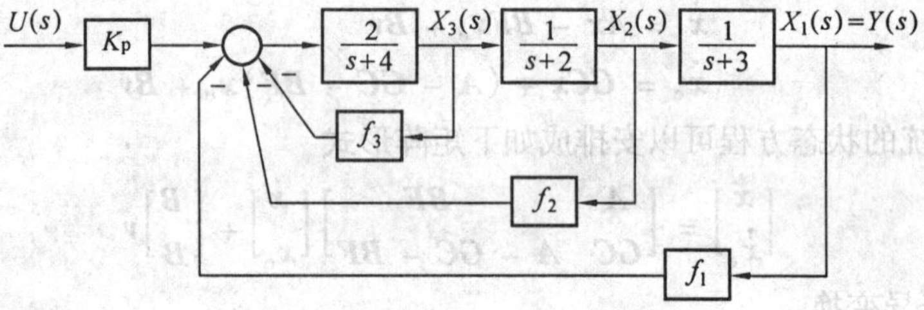

题 4.3 图　汽车悬架系统

题 4.4　考虑一个传递函数为 $\dfrac{(s-1)(s+2)}{(s+1)(s-2)(s+3)}$ 的系统，能够通过状态反馈使传递函数改变为 $\dfrac{(s-1)}{(s+2)(s+3)}$ 吗？如果可以，如何改变？

题 4.5　某系统的表达式为

$$\dot{x} = \begin{bmatrix} a & 1 \\ 0 & 0 \end{bmatrix} x + \begin{bmatrix} 0 \\ 1 \end{bmatrix} u,$$

$$y = \begin{bmatrix} 1 & 0 \end{bmatrix} x$$

设计一个状态观测器使得观测器的特征根位于 $-r$、$-2r(r > 0)$，并画出状态变量图。

题 4.6　某系统的状态空间表达式为

$$\dot{x} = \begin{bmatrix} 1 & 2 & 0 \\ 3 & -1 & 1 \\ 0 & 2 & 0 \end{bmatrix} x + \begin{bmatrix} 0 \\ 0 \\ 1 \end{bmatrix} u$$

$$y = [-1 \quad 1 \quad 1] x$$

设计一个降维观测器，观测器的所有特征根的位置都选在 -4。

题 4.7　某系统的状态空间表达式为

$$\dot{x} = \begin{bmatrix} 0 & 1 & 4 \\ 3 & 2 & 0 \\ 1 & 1 & 1 \end{bmatrix} x + \begin{bmatrix} 0 & 0 \\ 1 & 0 \\ 0 & 1 \end{bmatrix} u$$

$$y = \begin{bmatrix} 1 & 2 & 0 \\ 0 & 0 & 1 \end{bmatrix} x$$

(a) 设计一个降维观测器，使观测器的特征根位于 -5。

(b) 设计一个状态反馈控制器，使其闭环极点位于 -1、-2 和 -3。

(c) 画出带有降维观测器的状态反馈系统的状态变量图。

第5章　李雅普诺夫稳定性分析

稳定性是控制系统一个重要的定性性质,因为一个能够工作的系统必须是稳定的。在经典控制理论中,线性时不变系统的稳定性概念很容易理解。现在有必要把稳定性的概念扩展到非线性系统和时变系统。此外,还希望无需直接求解特征根就能确定稳定性。李雅普诺夫以他的第二法、或称为直接法处理这一问题。这一方法充分但并非必要的条件要求确定一个称之为李雅普诺夫函数的标量函数。李雅普诺夫法为稳定性问题提供了一种普遍适用的方法,以满足现代控制系统分析和综合的需要,通常这些系统是非线性的和(或)时变的,例如最优控制系统和自适应控制系统。

5.1　稳定性定义

5.1.1　平衡状态

没有强迫输入的系统可表示为

$$\dot{\boldsymbol{x}} = \boldsymbol{f}(\boldsymbol{x},t),\ \boldsymbol{x}(t_0) = \boldsymbol{x}_0 \tag{5.1}$$

式中 $\boldsymbol{x}$ 是一个 $n \times 1$ 状态向量,t_0 是初始时刻,而 $\boldsymbol{f}(\boldsymbol{x},t)$ 则是一个线性的或非线性的、时不变的或时变的 n 维函数,即

$$\dot{x}_i = f_i(x_1,x_2,\cdots,x_n;t),\ i = 1,2,\cdots,n \tag{5.2}$$

在稳定性问题的研究中,方程(5.1)的解通常记为 $\boldsymbol{x}(t;\boldsymbol{x}_0,t_0)$,以强调初始状态 $\boldsymbol{x}_0$ 和初始时间 t_0。就几何意义而言,响应 $\boldsymbol{x}(t;\boldsymbol{x}_0,t_0)$ 表示 n 维状态空间中在时刻 t_0 从初始状态 $\boldsymbol{x}_0$ 出发的一条轨迹。

在稳定性研究中有一个重要的想法,一个未施加输入的系统在其平衡状态附近的响应与系统的稳定性密切相关。如果对于任何 t_0 和所有 $t \geqslant t_0$ 都有

$$\dot{\boldsymbol{x}}_e = \boldsymbol{f}(\boldsymbol{x}_e,t) = \boldsymbol{0} \tag{5.3}$$

则称 $\boldsymbol{x}_e$ 是一个平衡状态。

由这一定义可以看到,如果状态空间中的一条状态轨迹到达一个平衡状态而且又没有输入施加于系统,这条轨迹将始终停留在这一平衡状态上。由此,$\dot{\boldsymbol{x}} = \boldsymbol{f}(\boldsymbol{x},t)$ 的平衡状态是 $\dot{\boldsymbol{x}} = \boldsymbol{f}(\boldsymbol{x},t) = \boldsymbol{0}$ 的解。显然,由于一个线性时不变系统 $\dot{\boldsymbol{x}} = \boldsymbol{A}\boldsymbol{x}$ 的平衡状态满足 $\boldsymbol{A}\boldsymbol{x}_e = \boldsymbol{0}$,如果 $\boldsymbol{A}$ 是非奇异的,则 $\dot{\boldsymbol{x}} = \boldsymbol{f}(\boldsymbol{x},t) = \boldsymbol{0}$ 存在唯一的零解 $\boldsymbol{x} = \boldsymbol{0}$,即存在一个位于状态空间原点的平衡状态。因此,在这种情况下一个系统的稳定性等价于平衡状态的稳定性。

至于非线性系统,平衡状态的数量取决于状态方程,因为 $\dot{\boldsymbol{x}} = \boldsymbol{f}(\boldsymbol{x},t) = \boldsymbol{0}$ 可以有不止一个解。

5.1.2　李雅普诺夫意义下的稳定性

对于任意的 t_0 和所有的 $t \geqslant t_0$，如果对于任意大于零的 ε 存在一个正数 $\delta(\varepsilon, t_0)$，使得任意的 $\| \boldsymbol{x}_0 - \boldsymbol{x}_e \| \leqslant \delta$ 都意味着

$$\| \boldsymbol{x}(t; x_0, t_0) - \boldsymbol{x}_e \| \leqslant \varepsilon \tag{5.4}$$

成立，则平衡状态 $\boldsymbol{x}_e$ 称为在李雅普诺夫意义下稳定的。在式(5.4) 中，符号 $\| \cdot \|$ 表示向量的范数。向量的范数是广义化的长度概念。粗略地讲，范数是状态空间中向量长度的度量。例如，这里的 $\| \boldsymbol{x}_0 - \boldsymbol{x}_e \|$ 用来表示状态空间中从点 $\boldsymbol{x}_0$ 到点 $\boldsymbol{x}_e$ 的“长度”，并定义

$$\| \boldsymbol{x}_0 - \boldsymbol{x}_e \| = \sqrt{(x_{10} - x_{1e})^2 + (x_{20} - x_{2e})^2 + \cdots + (x_{n0} - x_{ne})^2} \tag{5.5}$$

李雅普诺夫意义下的稳定性有着显而易见的几何意义。表达式 $\| \boldsymbol{x}_0 - \boldsymbol{x}_e \| \leqslant \delta$ 可以看成一个状态空间中半径为 δ、中心在平衡状态 $\boldsymbol{x}_e$ 的超球域，并记为 $S(\delta)$。与此相类似，表达式 $\| \boldsymbol{x}(t; x_0, t_0) - \boldsymbol{x}_e \| \leqslant \varepsilon$ 可以看成一个状态空间中半径为 ε、中心在平衡状态 $\boldsymbol{x}_e$ 的超球域，并记为 $S(\varepsilon)$。这样，李雅普诺夫意义下的稳定性意味着，对于所有的 $t \geqslant t_0$，从 $S(\delta)$ 内出发的任何轨迹都不会离开 $S(\varepsilon)$。

对于一个两维系统，上述几何意义可以如图 5.1(a) 所示，用状态空间中的一条轨迹加以说明。轨迹 ① 是闭合的，它表明是连续震荡；轨迹 ② 则终止于平衡状态。在这两种情况下，平衡状态 $\boldsymbol{x}_e$ 都是在李雅普诺夫意义下稳定的。

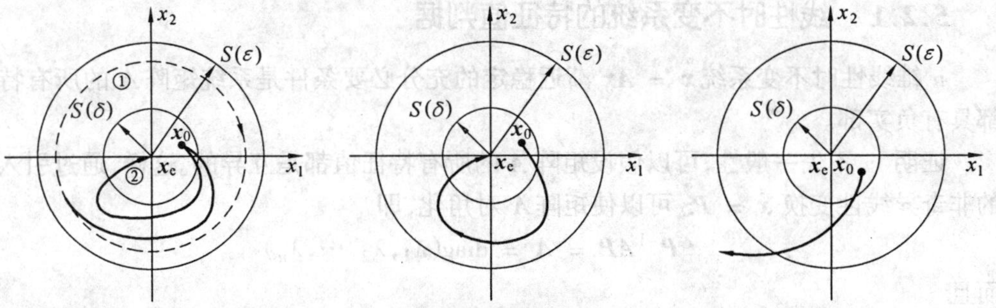

图 5.1　稳定性的几何表示

对于时变系统，δ 通常与 t_0 有关，而时不变系统的 δ 则与 t_0 无关。如果 δ 与 t_0 无关，那么这种稳定的平衡状态称为一致稳定的。

5.1.3　渐近稳定性

如果平衡状态 $\boldsymbol{x}_e$ 是在李雅普诺夫意义下稳定的，而且

$$\lim_{t \to \infty} \| \boldsymbol{x}(t; x_0, t_0) - \boldsymbol{x}_e \| \to 0 \tag{5.6}$$

则称此平衡状态是渐近稳定的。这意味着所有从 $S(\delta)$ 内出发的轨迹都不会离开 $S(\varepsilon)$，而且当时间 t 趋向无穷大时到达平衡状态。显然，这与通常在经典控制理论中所定义的稳定性是一致的。渐近稳定性的几何表示如图 5.1(b) 所示。

如果 δ 与 t_0 无关，那么平衡状态 $\boldsymbol{x}_e$ 称为一致渐近稳定的。

5.1.4 大范围稳定性

如果球域 $S(\delta)$ 可以扩展到整个状态空间,则稳定的平衡状态称为大范围稳定的。如果一个线性系统是稳定的,则必定是大范围稳定的,因为任何初始状态都会得到相同的稳定性结果。至于非线性系统,通常稳定性与初始条件密切有关,稳定的平衡状态只是在小范围内稳定。

如果 δ 与 t_0 无关,大范围稳定的平衡状态 $\boldsymbol{x}_e$ 称为一致大范围稳定的。

5.1.5 不稳定性

如果无论指定多么大的 ϵ 和多么小的 δ,只要存在从 $S(\delta)$ 出发的轨迹离开 $S(\epsilon)$,一个平衡状态就称为不稳定的。不稳定性的几何表示如图 5.1(c) 所示。显然,如果一个线性系统的平衡状态是不稳定的,则此系统是不稳定的。

5.2 李雅普诺夫第一法

李雅普诺夫第一法,即李雅普诺夫间接法是基于状态方程的解确定系统稳定性的方法。这一方法适用于线性时不变、线性时变和一些可线性化非线性函数的情况。

5.2.1 线性时不变系统的特征值判据

n 维线性时不变系统 $\dot{\boldsymbol{x}} = \boldsymbol{A}\boldsymbol{x}$ 渐近稳定的充分必要条件是系统矩阵 $\boldsymbol{A}$ 的所有特征值都具有负实部。

证明 不失一般性,可以假设矩阵 $\boldsymbol{A}$ 的所有特征值都是互异的。这样,通过引入合适的非奇异线性变换 $\boldsymbol{x} = \boldsymbol{P}\boldsymbol{z}$ 可以使矩阵 $\boldsymbol{A}$ 对角化,即

$$\boldsymbol{P}^{-1}\boldsymbol{A}\boldsymbol{P} = \boldsymbol{\Lambda} = \mathrm{diag}(\lambda_1, \lambda_2, \cdots, \lambda_n)$$

而且

$$\mathrm{e}^{\boldsymbol{A}t} = \boldsymbol{P}\mathrm{e}^{\boldsymbol{P}^{-1}\boldsymbol{A}\boldsymbol{P}t}\boldsymbol{P}^{-1} = \boldsymbol{P}\mathrm{e}^{\boldsymbol{\Lambda}t}\boldsymbol{P}^{-1}$$

将此式展开可以看到,$\mathrm{e}^{\boldsymbol{A}t}$ 的每一项都是 $\mathrm{e}^{\lambda_1 t}, \mathrm{e}^{\lambda_2 t}, \cdots, \mathrm{e}^{\lambda_n t}$ 的线性组合。因此 $\dot{\boldsymbol{x}} = \boldsymbol{A}\boldsymbol{x}$ 的解可以表示为

$$\boldsymbol{x}(t) = \mathrm{e}^{\boldsymbol{A}t}\boldsymbol{x}(0) = \boldsymbol{P}\mathrm{e}^{\boldsymbol{\Lambda}t}\boldsymbol{P}^{-1}\boldsymbol{x}(0) = [\boldsymbol{R}_1\mathrm{e}^{\lambda_1 t} + \boldsymbol{R}_2\mathrm{e}^{\lambda_2 t} + \cdots + \boldsymbol{R}_n\mathrm{e}^{\lambda_n t}\boldsymbol{P}^{-1}]\boldsymbol{x}(0)$$

其中 $\boldsymbol{R}_i, i = 1,2,\cdots,n$,是系数矩阵。很容易就能看到,系统渐近稳定,即对于任意给定的 $\boldsymbol{x}(0)$ 有 $\lim\limits_{t\to\infty}\boldsymbol{x}(t) \to 0$ 的充分必要条件是 $\mathrm{Re}(\lambda_i) < 0, i = 1,2,\cdots,n$。

当矩阵 $\boldsymbol{A}$ 具有重特征值时,在 $\boldsymbol{x}(t)$ 的表达式中将加入诸如 $t\mathrm{e}^{\lambda t}$、$t^2\mathrm{e}^{\lambda t}$ 之类的响应模式。在这种情况下,关于稳定性的结论并没有改变。

至于李雅普诺夫意义下的稳定性,如果矩阵 $\boldsymbol{A}$ 变换为约当形,则其充分必要条件为 $\boldsymbol{A}$ 的所有特征值都具有非正实部(负实部或零实部)且与零实部特征值相关的每一个约当块都是一阶的。

例 5.1 判断两个系统的稳定性,它们的系统矩阵分别为

(1) $\boldsymbol{A}_1 = \begin{bmatrix} 0 & 0 \\ 0 & 0 \end{bmatrix}$，(2) $\boldsymbol{A}_2 = \begin{bmatrix} 0 & 1 \\ 0 & 0 \end{bmatrix}$。

解　(1) 系统矩阵是约当形，而且两个特征值都为零。与特征值 $\lambda = 0$ 相关的约当块有两个，而且两个约当块的阶数都为 1。因此，该系统是在李雅普诺夫意义下稳定的，但它不是渐近稳定的。

(2) 系统矩阵也是约当形，而且两个特征值也都为零。由于与特征值 $\lambda = 0$ 相关的约当块的阶数为 2，该系统是不稳定的。

事实上，通过求取这两个系统的响应，可以很容易检验对于它们的结论。对于情况(1)，响应为

$$\boldsymbol{x}(t) = \mathrm{e}^{A_1 t}\boldsymbol{x}(0) = \begin{bmatrix} 1 & 0 \\ 0 & 1 \end{bmatrix}\begin{bmatrix} x_1(0) \\ x_2(0) \end{bmatrix} = \begin{bmatrix} x_1(0) \\ x_2(0) \end{bmatrix}$$

而在情况(2) 中，响应为

$$\boldsymbol{x}(t) = \mathrm{e}^{A_2 t}\boldsymbol{x}(0) = \begin{bmatrix} 1 & t \\ 0 & 1 \end{bmatrix}\begin{bmatrix} x_1(0) \\ x_2(0) \end{bmatrix} = \begin{bmatrix} x_1(0) + t x_2(0) \\ x_2(0) \end{bmatrix}$$

显然，第一个系统是且仅是在李雅普诺夫意义下稳定的，而第二个系统则是不稳定的。

5.2.2　线性化时不变系统的特征值判据

通常，一个非线性系统可以有不止一个平衡状态。当非线性系统可以在某一个平衡状态附近展开为泰勒级数时，在这一平衡状态的一个小邻域内该系统的行为就像线性系统一样，而且线性化系统的特征值可以用于确定这一平衡状态的稳定性。

假设非线性系统 $\dot{\boldsymbol{x}} = \boldsymbol{f}(\boldsymbol{x})$ 可以在平衡状态 $\boldsymbol{x}_\mathrm{e}$ 附近展开为以下的泰勒级数

$$\dot{\boldsymbol{x}} = \boldsymbol{f}(\boldsymbol{x}_\mathrm{e}) + \left.\frac{\partial \boldsymbol{f}}{\partial \boldsymbol{x}^\mathrm{T}}\right|_{\boldsymbol{x}=\boldsymbol{x}_\mathrm{e}} \cdot (\boldsymbol{x} - \boldsymbol{x}_\mathrm{e}) + \boldsymbol{g}(\boldsymbol{x}) \tag{5.7}$$

式中 $\boldsymbol{g}(\boldsymbol{x})$ 为泰勒级数中高次项之和。记

$$\Delta \boldsymbol{x} = \boldsymbol{x} - \boldsymbol{x}_\mathrm{e} \tag{5.8}$$

并忽略泰勒级数中的高次项可得到线性化的状态方程

$$\Delta\dot{\boldsymbol{x}} = \left.\frac{\partial \boldsymbol{f}}{\partial \boldsymbol{x}^\mathrm{T}}\right|_{\boldsymbol{x}=\boldsymbol{x}_\mathrm{e}} \cdot \Delta\boldsymbol{x} = \left.\begin{bmatrix} \dfrac{\partial f_1}{\partial x_1} & \dfrac{\partial f_1}{\partial x_2} & \cdots & \dfrac{\partial f_1}{\partial x_n} \\ \dfrac{\partial f_2}{\partial x_1} & \dfrac{\partial f_2}{\partial x_2} & \cdots & \dfrac{\partial f_2}{\partial x_n} \\ \vdots & \vdots & & \vdots \\ \dfrac{\partial f_n}{\partial x_1} & \dfrac{\partial f_n}{\partial x_2} & \cdots & \dfrac{\partial f_n}{\partial x_n} \end{bmatrix}\right|_{\boldsymbol{x}=\boldsymbol{x}_\mathrm{e}} \cdot \Delta\boldsymbol{x} = \boldsymbol{J}_x \Delta\boldsymbol{x} \tag{5.9}$$

式中 f_i 是 $\boldsymbol{f}(\boldsymbol{x})$ 的第 i 行，$\boldsymbol{J}_x = \partial \dfrac{\boldsymbol{f}^\mathrm{T}}{\partial \boldsymbol{x}}$ 称为雅可比矩阵，它是在 $\boldsymbol{x}_\mathrm{e}$ 处求得的。

对于如式(5.9) 线性化的非线性系统，如果 $\boldsymbol{J}_x$ 的全部特征值都具有负实部，则平衡状态 $\boldsymbol{x}_\mathrm{e}$ 是渐近稳定的；如果 $\boldsymbol{J}_x$ 的任一特征值具有正实部，则平衡状态 $\boldsymbol{x}_\mathrm{e}$ 是不稳定的。至于有些特征值具有零实部而另一些具有负实部的情况，则线性系统在 $\boldsymbol{x}_\mathrm{e}$ 处的稳定性与

$g(\boldsymbol{x})$ 有关。如果 $g(\boldsymbol{x}) = \boldsymbol{0}$，则平衡状态 $\boldsymbol{x}_e$ 就和线性系统的情况一样在李雅普诺夫意义下稳定或者不稳定，否则它的稳定性不能用李雅普诺夫第一法确定。

5.3 李雅普诺夫第二法

由机械系统的振动现象可以观察到，如果它的能量（包括动能和势能在内）随着时间的推移逐渐衰减，系统迟早都会到达平衡状态。由于难以为控制系统寻找到切实可行的能量函数表达式，李雅普诺夫提出了构造一个与状态变量有关的虚拟能量函数 $V(\boldsymbol{x},t)$，这一函数称为李雅普诺夫函数。由此，能量衰减的特性可以用 $\dot{V}(\boldsymbol{x},t)$ 表示。李雅普诺夫第二法，即李雅普诺夫直接法无须求解特征值，而是利用 $V(\boldsymbol{x},t)$ 和 $\dot{V}(\boldsymbol{x},t)$ 的符号特性即正定性确定系统的稳定性。这一方法可以应用于线性系统或非线性系统、时不变系统或时变系统及确定系统或随机系统。应用这一方法的主要困难是没有通用的方法构造合适的李雅普诺夫函数，因为某一个函数不能满足稳定性条件并不意味着没有真实存在的李雅普诺夫函数。

5.3.1 标量函数的正定性和西尔维斯特定理

1.标量函数的正定性

一个向量的标量函数，例如 $V(\boldsymbol{x})$，它的正定性是定义在用 $\|\boldsymbol{x}\| \leqslant K$ 描述的原点附近的一个球域 S 内，常数 K 为球域半径。

如果标量函数 $V(\boldsymbol{x})$ 在 S 域内对于 $\boldsymbol{x} = \boldsymbol{0}$ 有 $V(\boldsymbol{x}) = \boldsymbol{0}$，而且对于所有 $\boldsymbol{x} \neq \boldsymbol{0}$ 有 $V(\boldsymbol{x}) > 0$，则称 $V(\boldsymbol{x})$ 在 S 域内是正定的。例如，$V(\boldsymbol{x}) = x_1^2 + x_2^2$ 是正定的。

如果标量函数 $V(\boldsymbol{x})$ 在 S 域内对于 $\boldsymbol{x} = \boldsymbol{0}$ 有 $V(\boldsymbol{x}) = 0$，而且对于所有 $\boldsymbol{x} \neq \boldsymbol{0}$ 有 $V(\boldsymbol{x}) \geqslant 0$，则称 $V(\boldsymbol{x})$ 在 S 域内是正半定的。例如，$V(\boldsymbol{x}) = (x_1 + x_2)^2$ 是正半定的。

如果标量函数在 S 域内 $V(\boldsymbol{x})$ 对于 $\boldsymbol{x} = \boldsymbol{0}$ 有 $V(\boldsymbol{x}) = 0$，而且对于所有 $\boldsymbol{x} \neq \boldsymbol{0}$ 有 $V(\boldsymbol{x}) < 0$，则称 $V(\boldsymbol{x})$ 在 S 域内是负定的。如果 $-V(\boldsymbol{x})$ 是正定的，则 $V(\boldsymbol{x})$ 是负定的。例如，$V(\boldsymbol{x}) = -(x_1^2 + x_2^2)$ 是负定的。

如果标量函数 $V(\boldsymbol{x})$ 在 S 域内对于 $\boldsymbol{x} = \boldsymbol{0}$ 有 $V(\boldsymbol{x}) = 0$，而且对于所有 $\boldsymbol{x} \neq \boldsymbol{0}$ 有 $V(\boldsymbol{x}) \leqslant 0$，则称 $V(\boldsymbol{x})$ 在 S 域内是负半定的。如果 $-V(\boldsymbol{x})$ 是正半定的，则 $V(\boldsymbol{x})$ 是负半定的。例如，$V(\boldsymbol{x}) = -(x_1 + x_2)^2$ 是负半定的。

如果标量函数 $V(\boldsymbol{x})$ 在 S 域内对于 $\boldsymbol{x} \neq \boldsymbol{0}$ 可正可负，则称 $V(\boldsymbol{x})$ 在 S 域内是不定的。例如，$V(\boldsymbol{x}) = x_1x_2$ 是不定的。

对于时变的标量函数 $V(\boldsymbol{x},t)$，正定性的定义与时间有关。例如，如果 $t \geqslant t_0$ 时 $V(\boldsymbol{x})$ 在 S 域内对于 $\boldsymbol{x} = \boldsymbol{0}$ 有 $V(x,t) = 0$，而且对于所有 $\boldsymbol{x} \neq \boldsymbol{0}$ 有 $V(x,t) > 0$，则称 $V(\boldsymbol{x},t)$ 在 S 域内是正定的。

2.二次型函数

在用于确定线性系统的稳定性时，李雅普诺夫第二法采用表示成二次型的标量函数，这是一种很重要的标量函数。

二次型函数 $V(\boldsymbol{x})$ 是一个用实变量 $x_1, x_2, \cdots, x_n$ 表示的齐次多项式,其形式为

$$V(\boldsymbol{x}) = \sum_{i=1}^{n}\sum_{j=1}^{n} p_{ij} x_i x_j \tag{5.10}$$

其中所有的 p_{ij} 都是实数。这一齐次多项式可以记为

$$V(\boldsymbol{x}) = \boldsymbol{x}^{\mathrm{T}}\boldsymbol{P}\boldsymbol{x} = [x_1 \quad x_2 \quad \cdots \quad x_n]\begin{bmatrix} p_{11} & p_{12} & \cdots & p_{1n} \\ p_{21} & p_{22} & \cdots & p_{2n} \\ \vdots & \vdots & & \vdots \\ p_{n1} & p_{n2} & \cdots & p_{nn} \end{bmatrix}\begin{bmatrix} x_1 \\ x_2 \\ \vdots \\ x_n \end{bmatrix} \tag{5.11}$$

式中 $\boldsymbol{P}$ 是一个对称阵。例如

$$V(\boldsymbol{x}) = x_1^2 - 4x_2^2 + 5x_3^2 + 6x_1x_2 - 10x_2x_3 =$$

$$\boldsymbol{x}^{\mathrm{T}}\boldsymbol{P}\boldsymbol{x} = [x_1 \quad x_2 \quad x_3]\begin{bmatrix} 1 & 3 & 0 \\ 3 & -4 & -5 \\ 0 & -5 & 5 \end{bmatrix}\begin{bmatrix} x_1 \\ x_2 \\ x_3 \end{bmatrix}$$

3.西尔维斯特定理

二次型函数的正定性可以通过观察确定,也可以通过计算矩阵 $\boldsymbol{P}$ 的主子行列式来确定,后一种方法有时候称为西尔维斯特定理。在矩阵 $\boldsymbol{P}$ 中删去任意一些行和同序号的列,就可以得到它的一个主子行列式。矩阵 $\boldsymbol{P}$ 的顺序主子行列式则是它里面包含 p_{11} 的方阵列。一个 $n \times n$ 矩阵 $\boldsymbol{P}$ 有 n 个顺序主子行列式,它们由下式给出为

$$\Delta_1 = p_{11},\ \Delta_{12} = \begin{vmatrix} p_{11} & p_{12} \\ p_{21} & p_{22} \end{vmatrix},\ \Delta_{123} = \begin{vmatrix} p_{11} & p_{12} & p_{13} \\ p_{21} & p_{22} & p_{23} \\ p_{31} & p_{32} & p_{33} \end{vmatrix},\ \cdots,\ \Delta_{12\cdots n} = |\boldsymbol{P}|$$

Δ 的下标为用于组成主子行列式的行和列。

根据西尔维斯特定理,如果矩阵 $\boldsymbol{P}$ 的所有顺序主子行列式都大于零,则 $V(\boldsymbol{x})$ 是正定的,矩阵 $\boldsymbol{P}$ 也称为正定阵;如果 $|\boldsymbol{P}| = 0$ 且矩阵 $\boldsymbol{P}$ 的所有主子行列式都非负,则 $V(\boldsymbol{x})$ 是正半定的,矩阵 $\boldsymbol{P}$ 也称为正半定阵;如果 $-\boldsymbol{P}$ 是正定的,或者 $\boldsymbol{P}$ 的所有奇数阶顺序主子行列式都小于零而所有偶数阶顺序主子行列式都大于零,则 $V(\boldsymbol{x})$ 是负定的,矩阵 $\boldsymbol{P}$ 也称为负定阵;如果 $-\boldsymbol{P}$ 是正半定的,则 $V(\boldsymbol{x})$ 是负半定的,矩阵 $\boldsymbol{P}$ 也称为负半定阵;如果 $\boldsymbol{P}$ 不是正定的、正半定的、负定的,也不是负半定的,则 $V(\boldsymbol{x})$ 是不定的,$\boldsymbol{P}$ 也称为不定阵。

例 5.2　确定矩阵 $\boldsymbol{P}$ 的正定性,

$$\boldsymbol{P} = \begin{bmatrix} 1 & 1 & 1 \\ 1 & 1 & 1 \\ 1 & 1 & 0 \end{bmatrix}。$$

解　估算顺序主子行列式以检验它们的正定性

$$\Delta_1 = 1,\ \Delta_{12} = \begin{vmatrix} 1 & 1 \\ 1 & 1 \end{vmatrix} = 0,\ \Delta_{123} = \begin{vmatrix} 1 & 1 & 1 \\ 1 & 1 & 1 \\ 1 & 1 & 0 \end{vmatrix} = 0$$

检验结果不满足正定的条件。接着估算其他的主子行列式以检验其正定性

$$\Delta_{13} = \begin{vmatrix} 1 & 1 \\ 1 & 0 \end{vmatrix} = -1$$

由于主子行列式不全为非负，矩阵 $\boldsymbol{P}$ 不是正半定的。类似地可知，矩阵 $\boldsymbol{P}$ 也不是负定或负半定的。因此，矩阵 $\boldsymbol{P}$ 是不定阵。

5.3.2　李雅普诺夫稳定性理论

考虑非线性或线性系统

$$\dot{\boldsymbol{x}} = \boldsymbol{f}(\boldsymbol{x}, t) \tag{5.12}$$

不失一般性，假设平衡状态在状态空间的原点，即 $\boldsymbol{x}_e = \boldsymbol{0}$。

1.渐近稳定定理

对于由式(5.12) 描述的系统，如果存在一个具有连续一阶导数的标量函数 $V(\boldsymbol{x}, t)$，使得

(1) $V(\boldsymbol{x}, t)$ 是正定的；

(2) 以下条件之一得以满足：

①$\dot{V}(\boldsymbol{x}, t)$ 是负定的，或者

②$\dot{V}(\boldsymbol{x}, t)$ 是负半定的，而且对任意 $\boldsymbol{x} \neq \boldsymbol{0}$ 有 $\dot{V}[\boldsymbol{x}(t; \boldsymbol{x}_0, t_0)]$ 不恒为零；

则平衡状态 $\boldsymbol{x}_e = 0$是渐近稳定的，$V(\boldsymbol{x}, t)$则称为式(5.12)系统的李雅普诺夫函数。而且，如果 $\|\boldsymbol{x}\| \to \infty$ 时 $V(\boldsymbol{x}, t) \to \infty$，平衡状态 $\boldsymbol{x}_e = 0$ 是大范围渐近稳定的。

这一定理可以用一个二阶系统予以说明。函数 $V(\boldsymbol{x}) = x_1^2 + x_2^2$是正定的，并可以用图5.2(a) 所示椭球面表示。$V(\boldsymbol{x}) = c$ 用椭球面 $V(\boldsymbol{x})$ 与平面 $z = c$ 相交的轨迹表示，c 是常数。这一轨迹在$[x_1, x_2]$ 平面上的投影是一条围绕原点的椭圆形闭合曲线。对于不同的 c 值，在$[x_1, x_2]$平面上就形成这样一簇闭合曲线。如果 $\dot{V}(\boldsymbol{x})$负定，起始于初始状态 $\boldsymbol{x}_0$ 的响应轨迹将接连不断地穿过 $V(\boldsymbol{x})$ 值越来越小的椭圆，最终当 $t \to \infty$ 时到达原点，如图 5.2(b) 所示。如果 $\dot{V}(\boldsymbol{x})$ 是负半定的，起始于初始状态 $\boldsymbol{x}_0$ 的响应轨迹将与某一椭圆相切，在切点处 $\dot{V}(\boldsymbol{x}) = 0$。但是，由于 $\dot{V}(\boldsymbol{x})$ 对于任何 $\boldsymbol{x} \neq 0$ 不恒为零，响应轨迹不会停留在切点，而是继续向着原点运动，如图 5.2 (c) 所示。

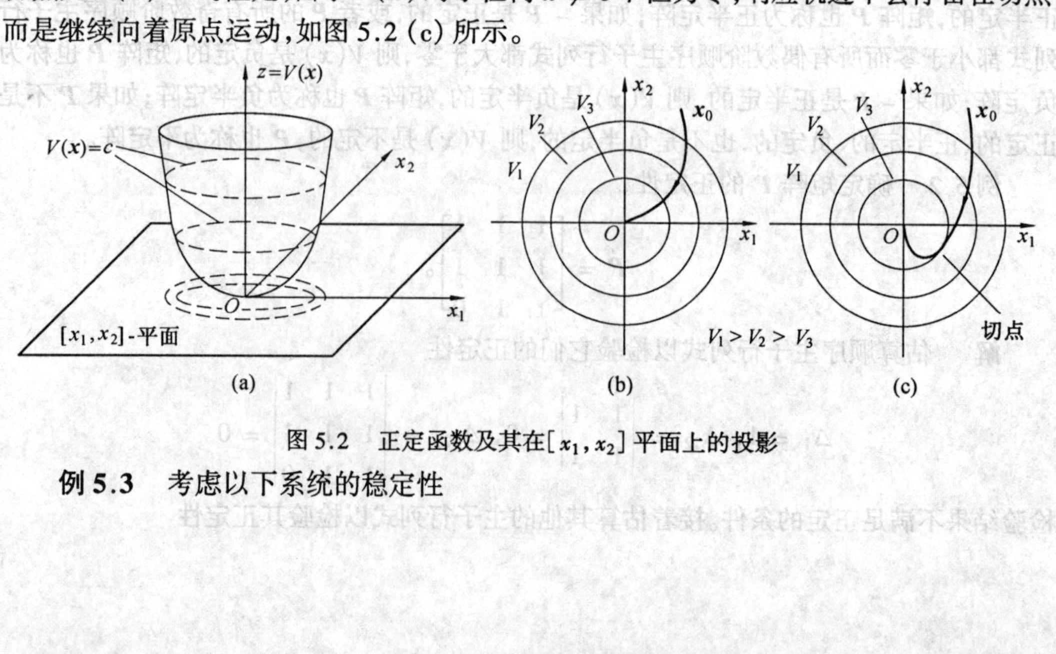

图 5.2　正定函数及其在$[x_1, x_2]$ 平面上的投影

例 5.3　考虑以下系统的稳定性

$$\dot{x}_1 = x_2 - x_1(x_1^2 + x_2^2)$$
$$\dot{x}_2 = -x_1 - x_2(x_1^2 + x_2^2)$$

解　令 $\dot{x}_1 = 0$ 和 $\dot{x}_2 = 0$ 得到解 $x_1 = 0$ 和 $x_2 = 0$，即原点是该系统唯一的平衡状态。一个简单的正定二次型标量函数是

$$V(\boldsymbol{x}) = x_1^2 + x_2^2$$

它的导数为

$$\dot{V}(\boldsymbol{x}) = 2x_1\dot{x}_1 + 2x_2\dot{x}_2 = -2(x_1^2 + x_2^2)^2$$

函数 $\dot{V}(\boldsymbol{x})$ 是负定的，这意味着 $V(\boldsymbol{x})$ 沿着轨迹连续不断地减小。因此，$\boldsymbol{x}_e = \boldsymbol{0}$ 是渐近稳定的，而且 $V(\boldsymbol{x})$ 是给定系统的一个李雅普诺夫函数。由于 $\|\boldsymbol{x}\| \to \infty$ 时 $V(\boldsymbol{x}) \to \infty$，$\boldsymbol{x}_e = \boldsymbol{0}$ 是大范围渐近稳定的。

例 5.4　考虑以下系统的稳定性

$$\dot{x}_1 = x_2$$
$$\dot{x}_2 = -x_1 - x_2$$

解　显然，原点 $\boldsymbol{x} = \boldsymbol{0}$ 是系统唯一的平衡状态。令

$$V(\boldsymbol{x}) = 2x_1^2 + x_2^2$$

得到

$$\dot{V}(\boldsymbol{x}) = 4x_1\dot{x}_1 + 2x_2\dot{x}_2 = 2x_1x_2 - 2x_2^2$$

$V(\boldsymbol{x})$ 正定的，但 $\dot{V}(\boldsymbol{x})$ 是不定的。应当注意，这一结果并不意味着没有确实存在的李雅普诺夫函数。如果选择

$$V(\boldsymbol{x}) = x_1^2 + x_2^2$$

则有

$$\dot{V}(\boldsymbol{x}) = 2x_1\dot{x}_1 + 2x_2\dot{x}_2 = -2x_2^2$$

现在 $V(\boldsymbol{x})$ 是正定的，且 $\dot{V}(\boldsymbol{x})$ 是正半定的。在这种情况下，由必要研究 $\dot{V}(\boldsymbol{x})$ 对于任何 $\boldsymbol{x} \neq \boldsymbol{0}$ 是否恒为零。令 $\dot{V}(\boldsymbol{x}) \equiv 0$ 有 $x_2 \equiv 0$，这要求 $\dot{x}_2 = 0$。但是，由描述该系统的方程可知，这一要求意味着 x_1 必须为零。因此，$\dot{V}(\boldsymbol{x})$ 对于任何 $\boldsymbol{x} \neq \boldsymbol{0}$ 不恒为零。所以，$\boldsymbol{x}_e = \boldsymbol{0}$ 是渐近稳定的，所选的 $V(\boldsymbol{x})$ 是一个李雅普诺夫函数。而且，由于 $\|\boldsymbol{x}\| \to \infty$ 时 $V(\boldsymbol{x}) \to \infty$，$\boldsymbol{x}_e = \boldsymbol{0}$ 是大范围渐近稳定的。

还应当注意的是，李雅普诺夫函数不是唯一的，例如 $V(\boldsymbol{x}) = [(x_1 + x_2)^2 + 2x_1^2 + x_2^2]/2$ 是例 5.4 系统的另一个李雅普诺夫函数。

2. 李雅普诺夫意义下稳定定理

对于式(5.12)所描述的系统，如果存在具有连续一阶导数的标量函数 $V(\boldsymbol{x}, t)$ 使得

(1) $V(\boldsymbol{x}, t)$ 是正定的；

(2) $\dot{V}(\boldsymbol{x}, t)$ 是负半定的，且 $\dot{V}[\boldsymbol{x}(t; \boldsymbol{x}_0, t_0)]$ 对于有些 $\boldsymbol{x} \neq \boldsymbol{0}$ 恒为零；则 $\boldsymbol{x}_e = \boldsymbol{0}$ 是在李雅普诺夫意义下稳定的，且 $V(\boldsymbol{x}, t)$ 是式(5.12)系统的一个李雅普诺夫函数。

例 5.5　考虑下面系统的稳定性

$$\dot{x}_1 = kx_2$$
$$\dot{x}_2 = -x_1$$

其中 $k > 0$。

解　显然,给定系统在原点 $\boldsymbol{x} = \boldsymbol{0}$ 有一个唯一的平衡状态。选择一个正定的标量函数

$$V(\boldsymbol{x}) = x_1^2 + kx_2^2$$

则有

$$\dot{V}(\boldsymbol{x}) = 2x_1\dot{x}_1 + 2kx_2\dot{x}_2 = 2kx_1x_2 - 2kx_1x_2 = 0$$

可以看到,对于任何 $\boldsymbol{x} \neq \boldsymbol{0}$ 都有 $\dot{V}(\boldsymbol{x})$ 恒为零。因此,该系统是在李雅普诺夫意义下稳定的,但不是渐近稳定的。

3. 不稳定定理

对于式(5.12)所描述的系统,如果存在具有连续一阶导数的标量函数 $V(\boldsymbol{x}, t)$ 使得

(1) $V(\boldsymbol{x}, t)$ 是正定的;

(2) 下列条件之一得到满足:

①$\dot{V}(\boldsymbol{x}, t)$ 是正定的,或者

②$\dot{V}(\boldsymbol{x}, t)$ 是正半定的,且 $\dot{V}[\boldsymbol{x}(t; \boldsymbol{x}_0, t_0)]$ 对于任一 $\boldsymbol{x} \neq \boldsymbol{0}$ 不恒为零,则 $\boldsymbol{x}_e = \boldsymbol{0}$ 是不稳定的。

例 5.6　判断下面系统的稳定性

$$\dot{x}_1 = x_1 + x_2$$
$$\dot{x}_2 = -x_1 + x_2$$

解　给定系统在原点 $\boldsymbol{x} = \boldsymbol{0}$ 处有一个唯一的平衡状态。选择一个正定的标量函数

$$V(\boldsymbol{x}) = x_1^2 + x_2^2$$

则有

$$\dot{V}(\boldsymbol{x}) = 2x_1\dot{x}_1 + 2x_2\dot{x}_2 = 2x_1^2 + 2x_2^2$$

通过观察可知,$\dot{V}(\boldsymbol{x})$ 是正定的,$\boldsymbol{x}_e = \boldsymbol{0}$ 是不稳定的。

例 5.7　考虑下面系统的稳定性

$$\dot{x}_1 = x_2$$
$$\dot{x}_2 = -x_1 + x_2$$

解　给定系统也是在原点 $\boldsymbol{x} = \boldsymbol{0}$ 处有一个唯一的平衡状态。选择一个正定的标量函数

$$V(\boldsymbol{x}) = x_1^2 + x_2^2$$

则有

$$\dot{V}(\boldsymbol{x}) = 2x_1\dot{x}_1 + 2x_2\dot{x}_2 = 2x_2^2$$

可以看到,$\dot{V}(\boldsymbol{x})$ 是负半定的。令 $\dot{V}(\boldsymbol{x}) \equiv \boldsymbol{0}$ 则有 $x_2 \equiv 0$,这要求 $\dot{x}_2 = 0$。但是,有系统的状态方程,何以要求意味着 x_1 必须为零。这说明 $\dot{V}(\boldsymbol{x})$ 对于任何 $\boldsymbol{x} \neq \boldsymbol{0}$ 不恒为零。所以 $\boldsymbol{x}_e = \boldsymbol{0}$ 是不稳定的。

5.4　线性系统的李雅普诺夫法应用

李雅普诺夫第二法可以应用于时变系统和非线性系统。尽管仍然没有简单而通用的

方法获取李雅普诺夫函数 $V(\boldsymbol{x},t)$,对于许多这一类的系统,已经逐渐研究出了解决问题的方法。尤其是对于线性系统而言,可以用以下方法确定李雅普诺夫函数 $V(\boldsymbol{x},t)$。

5.4.1　时不变连续系统

考虑由以下状态方程表示的 n 维无输入作用系统

$$\dot{\boldsymbol{x}}(t) = \boldsymbol{A}\boldsymbol{x} \tag{5.13}$$

其中 $\boldsymbol{A}$ 是非奇异的。显然,该系统在原点处有一个唯一的平衡状态。可能的李雅普诺夫函数 $V(\boldsymbol{x})$ 用对称阵 $\boldsymbol{P}$ 表示成

$$V(\boldsymbol{x}) = \boldsymbol{x}^{\mathrm{T}}\boldsymbol{P}\boldsymbol{x} \tag{5.14}$$

$V(\boldsymbol{x})$ 的导数为

$$\dot{V}(\boldsymbol{x}) = \dot{\boldsymbol{x}}^{\mathrm{T}}\boldsymbol{P}\boldsymbol{x} + \boldsymbol{x}^{\mathrm{T}}\boldsymbol{P}\dot{\boldsymbol{x}} = \boldsymbol{x}^{\mathrm{T}}(\boldsymbol{A}^{\mathrm{T}}\boldsymbol{P} + \boldsymbol{P}\boldsymbol{A})\boldsymbol{x} \tag{5.15}$$

令

$$\boldsymbol{A}^{\mathrm{T}}\boldsymbol{P} + \boldsymbol{P}\boldsymbol{A} = -\boldsymbol{Q} \tag{5.16}$$

则有

$$\dot{V}(\boldsymbol{x}) = -\boldsymbol{x}^{\mathrm{T}}\boldsymbol{Q}\boldsymbol{x} \tag{5.17}$$

式(5.16) 称为李雅普诺夫方程。可以证明,对于式(5.13) 给出的系统,原点处平衡状态渐近稳定的充分必要条件是:对于任意给定的正定实对称阵 $\boldsymbol{Q}$,存在唯一的正定实对称阵 $\boldsymbol{P}$ 满足式(5.16) 中的李雅普诺夫方程。而且,$\boldsymbol{x}^{\mathrm{T}}\boldsymbol{P}\boldsymbol{x}$ 就是该系统的一个李雅普诺夫函数。

根据这一结论,系统的稳定性可以按以下步骤确定。

第 1 步　任意选择一个正定或正半定的 $n \times n$ 对称阵 $\boldsymbol{Q}$ 。例如,为了方便起见,像 $\boldsymbol{Q} = \boldsymbol{I}$ 这样简单的对角阵就是正定的。也可以选择正半定 $\boldsymbol{Q}$ 阵,在主对角线上除了任一元素为正数外其他元素全都为零。

第 2 步　使李雅普诺夫方程的各项相等确定 $\boldsymbol{P}$ 阵的元素。

第 3 步　利用西尔维斯特定理确定 $\boldsymbol{P}$ 阵的正定性。

第 4 步　如果 $\boldsymbol{P}$ 是正定的,则原点处的平衡状态是渐近稳定的,如果 $\boldsymbol{P}$ 是负定的,则原点处的平衡状态是不稳定的。

例 5.8　应用李雅普诺夫方程确定系统的稳定性

$$\dot{\boldsymbol{x}} = \begin{bmatrix} 0 & 1 \\ -2 & -3 \end{bmatrix}\boldsymbol{x}$$

解　将对称阵 $\boldsymbol{P}$ 记为

$$\boldsymbol{P} = \begin{bmatrix} p_{11} & p_{12} \\ p_{12} & p_{22} \end{bmatrix}$$

并选择 $\boldsymbol{Q} = \boldsymbol{I}$。由李雅普诺夫方程有

$$\begin{bmatrix} 0 & -2 \\ 1 & -3 \end{bmatrix}\begin{bmatrix} p_{11} & p_{12} \\ p_{12} & p_{22} \end{bmatrix} + \begin{bmatrix} p_{11} & p_{12} \\ p_{12} & p_{22} \end{bmatrix}\begin{bmatrix} 0 & 1 \\ -2 & -3 \end{bmatrix} = \begin{bmatrix} -1 & 0 \\ 0 & -1 \end{bmatrix}$$

由这一等式我们可以得到三个联立方程

$$\begin{cases} -4p_{12}=0 \\ -2p_{22}+p_{11}-3p_{12}=0 \\ 2p_{12}-6p_{22}=-1 \end{cases}$$

解得矩阵为

$$P=\begin{bmatrix} 5/4 & 1/4 \\ 1/4 & 1/4 \end{bmatrix}$$

由于 P 阵是正定的，该系统渐近稳定。而且，可以获得一个李雅普诺夫函数为

$$V(x)=X^{\mathrm{T}}PX=\frac{5}{4}x_1^2+\frac{1}{2}x_1x_2+\frac{1}{4}x_2^2$$

例 5.9 确定 k 的取值范围，使得图 5.3 所示系统渐近稳定。

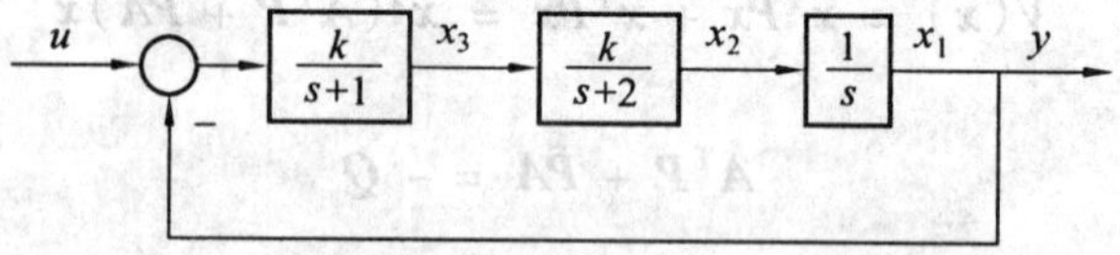

图 5.3 例 5.9 的方框图

解 假设指定状态变量如方框图中所示，则状态方程由下式给出

$$\dot{x}=\begin{bmatrix} 0 & 1 & 0 \\ 0 & -2 & 1 \\ -k & 0 & -1 \end{bmatrix}x+\begin{bmatrix} 0 \\ 0 \\ k \end{bmatrix}u$$

为了研究稳定性令 $u=0$，由于 $\det A=-k\neq 0$，系统矩阵非奇异，原点是该系统唯一的平衡状态。如果矩阵 Q 选为正半定的

$$Q=\begin{bmatrix} 0 & 0 & 0 \\ 0 & 0 & 0 \\ 0 & 0 & 1 \end{bmatrix}$$

则 $\dot{V}(x)=-x_3^2$ 是负半定的。令 $\dot{V}(x)\equiv 0$ 得到 $x_3\equiv 0$。由此，考虑到状态方程，由 $\dot{x}_3=-kx_1+x_3$ 有 $x_1\equiv 0$，由 $\dot{x}_1=x_2$ 则有 $x_2\equiv 0$。这表明采用这一正半定的矩阵 Q 可以使稳定性的分析得到简化。

李雅普诺夫方程为

$$\begin{bmatrix} 0 & 0 & -k \\ 1 & -2 & 0 \\ 0 & 1 & -1 \end{bmatrix}\begin{bmatrix} p_{11} & p_{12} & p_{13} \\ p_{12} & p_{22} & p_{23} \\ p_{13} & p_{23} & p_{33} \end{bmatrix}+\begin{bmatrix} p_{11} & p_{12} & p_{13} \\ p_{12} & p_{22} & p_{23} \\ p_{13} & p_{23} & p_{33} \end{bmatrix}\begin{bmatrix} 0 & 1 & 0 \\ 0 & -2 & 1 \\ -k & 0 & -1 \end{bmatrix}=\begin{bmatrix} 0 & 0 & 0 \\ 0 & 0 & 0 \\ 0 & 0 & -1 \end{bmatrix}$$

最终得到的解为

$$P=\frac{1}{12-2k}\begin{bmatrix} k^2+12k & 6k & 0 \\ 6k & 3k & k \\ 0 & k & 6 \end{bmatrix}$$

由西尔维斯特定理，矩阵 P 正定的充分必要条件是 $0<k<6$。因此，当 $0<k<6$ 时，该系统渐近稳定。

5.4.2　时不变离散系统

考虑用下面状态方程所表示的时不变离散系统

$$\boldsymbol{x}(k+1)=\boldsymbol{G}\boldsymbol{x}(k) \tag{5.18}$$

其中 $\boldsymbol{G}$ 是非奇异的。该系统在原点有一个唯一的平衡状态。选择正定的李雅普诺夫函数 $V[\boldsymbol{x}(k)]$，用对称阵 $\boldsymbol{P}$ 表示为

$$V[\boldsymbol{x}(k)]=\boldsymbol{x}^{\mathrm{T}}(k)\boldsymbol{P}\boldsymbol{x}(k) \tag{5.19}$$

$V[\boldsymbol{x}(k)]$ 的差分为

$$\begin{aligned}\Delta V[\boldsymbol{x}(k)] &= V[\boldsymbol{x}(k+1)]-V[\boldsymbol{x}(k)]=\boldsymbol{x}^{\mathrm{T}}(k+1)\boldsymbol{P}\boldsymbol{x}(k+1)-\boldsymbol{x}^{\mathrm{T}}(k)\boldsymbol{P}\boldsymbol{x}(k)=\\ &[\boldsymbol{G}\boldsymbol{x}(k)]^{\mathrm{T}}\boldsymbol{P}[\boldsymbol{G}\boldsymbol{x}(k)]-\boldsymbol{x}^{\mathrm{T}}(k)\boldsymbol{P}\boldsymbol{x}(k)=\\ &\boldsymbol{x}^{\mathrm{T}}(k)[\boldsymbol{G}^{\mathrm{T}}\boldsymbol{P}\boldsymbol{G}-\boldsymbol{P}]\boldsymbol{x}(k)\end{aligned} \tag{5.20}$$

令

$$\boldsymbol{G}^{\mathrm{T}}\boldsymbol{P}\boldsymbol{G}-\boldsymbol{P}=-\boldsymbol{Q} \tag{5.21}$$

得到

$$\Delta V[\boldsymbol{x}(k)]=-\boldsymbol{x}^{\mathrm{T}}(k)\boldsymbol{Q}\boldsymbol{x}(k) \tag{5.22}$$

式 (5.21) 称为离散李雅普诺夫方程。

可以证明，式(5.18) 给定的离散系统渐近稳定的充分必要条件为：对于任意给定的正定实对称矩阵 $\boldsymbol{Q}$，存在一个唯一的正定实对称矩阵 $\boldsymbol{P}$，满足式(5.21) 中的李雅普诺夫方程。

习　题

题 5.1　用雅可比矩阵确定用以下系统在平衡点附近的稳定性，描述系统的微分方程为

(a) $\begin{cases}\dot{x}_1 = x_1 + x_1^2 - x_1x_2\\ \dot{x}_2 = x_1 - x_2\end{cases}$

(b) $\begin{cases}\dot{x}_1 = 4x_1 - x_1x_2\\ \dot{x}_2 = x_1^2 - x_2\end{cases}$

(c) $\begin{cases}\dot{x}_1 = -x_1^2 + x_2\\ \dot{x}_2 = 4x_1 - 2x_2\end{cases}$

题 5.2　对于由以下方程描述的系统

(1) $\begin{cases}\dot{x}_1 = x_2\\ \dot{x}_2 = -x_1 - 3x_2 + 0.25x_3^2\\ \dot{x}_3 = x_1x_3 - x_3\end{cases}$

(2) $\begin{cases}\dot{x}_1 = x_2\\ \dot{x}_2 = -x_2 - 2\sin x_1\end{cases}$

(a) 确定各系统的平衡点。

(b) 求取关于平衡点的线性化方程。

(c) 确定系统在其平衡点附近是渐近稳定还是不稳定的。

题 5.3　写出以下二次型函数的向量 - 矩阵形式，并用西尔维斯特判据确定它们的正定性。

(a) $V(\boldsymbol{x}) = x_1^2 + 2x_2^2$

(b) $V(\boldsymbol{x}) = -x_1^2 - (3x_1^2 + 2x_2^2)^2$

(c) $V(\boldsymbol{x}) = -(x_1 + 2x_2)^2$

(d) $V(\boldsymbol{x}) = (x_1 + x_2)^2$

(e) $V(\boldsymbol{x}) = x_1x_2 + x_2^2$

(f) $V(\boldsymbol{x}) = -x_1^2 - 3x_2^2 - 11x_3^2 + 2x_1x_2 - 4x_2x_3 - 2x_1x_3$

(g) $V(\boldsymbol{x}) = x_1^2 + 4x_2^2 + x_3^2 + 2x_1x_2 - 6x_2x_3 - 2x_1x_3$

题 5.4　用李雅普诺夫第二法确定以下系统的稳定性。

(a) $\dot{\boldsymbol{x}} = \begin{bmatrix} -1 & 1 \\ 2 & -3 \end{bmatrix}\boldsymbol{x}$

(b) $\dot{\boldsymbol{x}} = \begin{bmatrix} 2 & 0.5 & -3 \\ 0 & -1 & 0 \\ 0 & 0.5 & -1 \end{bmatrix}$

题 5.5　用李雅普诺夫方程确定使得李雅普诺夫方程 $V(\boldsymbol{x}) = \boldsymbol{x}^{\mathrm{T}}\boldsymbol{P}\boldsymbol{x}$ 中得到正定阵 $\boldsymbol{P}$，从而确保以下系统渐近稳定时 K 的取值范围。

(a) $\boldsymbol{A} = \begin{bmatrix} -1 & K \\ 0 & -2 \end{bmatrix}$, $\boldsymbol{Q} = \begin{bmatrix} 0 & 0 \\ 0 & 1 \end{bmatrix}$

(b) $\boldsymbol{A} = \begin{bmatrix} -1 & 0 & 1 \\ 2 & -2 & 0 \\ K & 0 & -2 \end{bmatrix}$, $\boldsymbol{Q} = \begin{bmatrix} 0 & 0 & 0 \\ 0 & 1 & 0 \\ 0 & 0 & 0 \end{bmatrix}$

题 5.6　用李雅普诺夫方程确定使得李雅普诺夫函数 $V[\boldsymbol{x}(k)] = \boldsymbol{x}^{\mathrm{T}}(k)\boldsymbol{P}\boldsymbol{x}(k)$ 中得到正定阵 $\boldsymbol{P}$，从而确保下式渐近稳定时 K 的取值范围。

$$\boldsymbol{A} = \begin{bmatrix} 0 & 1 & 0 \\ 0 & 0 & 1 \\ 0 & \dfrac{K}{2} & 10 \end{bmatrix}, \quad \boldsymbol{Q} = \begin{bmatrix} 0 & 0 & 0 \\ 0 & 0 & 0 \\ 0 & 0 & 1 \end{bmatrix}$$

参考文献

[1] 多尔夫,等.现代控制系统(英文影印版)[M].北京:科学出版社,2002.
[2] 德赖斯.线性控制系统工程(英文影印版)[M].北京:清华大学出版社,2002.
[3] J'AZZO J, H HOUPIS C. Linear Control System Analysis and Design [M]. New York: McGraw-Hill, 1988.
[4] 郑大钟.线性系统理论[M].北京:清华大学出版社,2002.
[5] 李友善.自动控制原理[M].北京:国防工业出版社,1989.
[6] 李文秀.自动控制原理[M].哈尔滨:哈尔滨工程大学出版社,2001.
[7] 胡寿松.自动控制原理[M].北京:科学出版社,2002.
[8] 陈启宗.线性系统理论与设计[M].北京:科学出版社,1988.
[9] 王诗宓,等.自动控制理论例题习题集[M].北京:清华大学出版社,2002.